MW00620173

PILLARS OF CREATION

ALSO BY RICHARD PANEK

The Trouble with Gravity:
Solving the Mystery Beneath Our Feet

The 4% Universe:
Dark Matter, Dark Energy,
and the Race to Discover the Rest of Reality

The Invisible Century:
Einstein, Freud, and the Search for Hidden Universes

Seeing and Believing:
How the Telescope Opened
Our Eyes and Minds to the Heavens

Waterloo Diamonds:
A Midwestern Town and Its Minor League Team

BY TEMPLE GRANDIN AND RICHARD PANEK

The Autistic Brain:
Helping Different Kinds of Minds Succeed

PILLARS OF CREATION

HOW THE JAMES WEBB TELESCOPE UNLOCKED THE SECRETS OF THE COSMOS

RICHARD PANEK

LITTLE, BROWN AND COMPANY
New York Boston London

Little, Brown and Company
Hachette Book Group
1290 Avenue of the Americas, New York, NY 10104
littlebrown.com

First Edition: October 2024

For photo and illustration credits, see page 227.

ISBN 9780316570695
LCCN 2024934969

Printing 1, 2024

LSC-C

Printed in the United States of America

To Meg,
with love

I’m looking at the river
But I’m thinking of the sea

— *Randy Newman*

CONTENTS

Contents

GUIDE TO THE COLOR PHOTOGRAPHS

Page 1

The Webb Deep Field: This image was the public's initial glimpse of the power of Webb. The eight spiked objects are foreground stars. The rest are galaxies, some dating to the first billion years of the universe's existence. For more on the decades of preparation leading to Webb's scientific commissioning, see the Prologue and Chapters One and Two. For more on the initial release of Webb images in particular, see pages 9 and 84.

Pages 2–3

By Jupiter: This image of the gas giant vividly reveals several features of not just the planet itself but its immediate vicinity. The planet's aurorae are visible at its north and south poles but, outside the planet's disk, so is the auroral diffraction of light. The planet's rings, though faint, are evident to its left and right. Also to the left are two of Jupiter's moons: Adrastea at the tip of the rings and, farther to the left, Amalthea. (Jupiter's Great Red Spot, a storm that spans more than twice the width of the Earth, looks white because of the choice of filters and colors.) Upon closer examinations of Webb data, astronomers discovered a high-speed jet stream more than 4,800 kilometers (3,000 miles) wide above the clouds at the planet's equator. For more on astronomy within our solar system and Webb's contributions to that field, see Chapter Three. For more on Jupiter in particular, see pages 107–108.

Pages 4–5

A star is born: This image of the creation of a star system about 1,000 light-years from Earth reveals a possible infant analog to our Sun and its planetary system. The violent processes at the heart of star birth can produce stellar winds or jets of gas that collide with nearby gas and dust, creating dramatic phenomena such as the bow shocks in this portrait. For more on astronomy within our galaxy and Webb's contributions to that field, see Chapter Four. For more on protostars in particular, see pages 127–128.

Pages 6–7

Cosmic Cliffs: Another of the initial release of Webb images, this composite photograph shows a part of the Carina Nebula that is an example of what astronomers call a "stellar factory," a region of hyperkinetic star creation. This factory is 7,600 light-years from Earth. For more on the genesis of stars within our galaxy, see pages 160–161.

Page 8

Pillars of Creation, Part One: The Hubble Space Telescope's 1995 image of this stellar factory, about 6,500 light-years from Earth, popularized the term "Pillars of Creation." In returning to this region with Webb, astronomers chose to observe through several filters corresponding to specific wavelengths on the electromagnetic spectrum, to which technicians and astronomers assigned colors. This page and opposite: For more on the art and science of Webb images, see pages 138–142.

Page 9

Pillars of Creation, Part Two: The public release of the image reflects a composite of six filters plus further refinements for aesthetic and scientific purposes. Former protostars that have reached the hydrogen fusion stage—and thereby have gained the official status of "stars"—are visible as bright red dots at the "fingertip" ends of the pillars, lending the image a distinctly E.T. vibe.

Pages 10–11

Dust, dust, everywhere: "Empty" space isn't actually empty. It's full of dust and other ejecta of the potentially star-, planet-, and even life-forming variety, as revealed in this image of galaxy NGC 6822, 1.5 million light-years from Earth. For more on astronomy within the galaxies beyond our own Milky Way and Webb's contributions to that field, see Chapter Five. For more on the relationship between galactic evolution and dust, see pages 154–164.

Pages 12–13

The many faces of spiral galaxies: The "Physics at High Angular resolution in Nearby GalaxieS" (PHANGS) project used Webb to follow up its previous observations (some via the Hubble Space Telescope) of nineteen galaxies, all face-on spirals but each exhibiting a unique structure. For more on PHANGS, see pages 162–164.

Page 14

Stephan's Quintet: Another of Webb's initial releases, this mosaic consists of 150 million pixels and 1,000 image files via Webb's Near-Infrared Camera and Mid-Infrared Instrument. The galaxy on the left is about 40 million light-years from Earth, while the other four are about 290 million light-years distant. Moviegoers might recognize this grouping of galaxies from the opening of *It's a Wonderful Life* as the backdrop for a conversation between God and angels. This page and opposite: For more on the evolution of galaxies, see pages 159–162.

Page 15

Gravity rules: As with everything else in the universe, galaxies interact gravitationally, sometimes to spectacular effect. This pair of merging galaxies lies roughly 500 million light-years from Earth in the constellation Delphinus.

Page 16

A trick of the light: Astronomers use gravitational lensing—an effect predicted by Einstein's general theory of relativity—to study objects in the early universe. The great mass of a foreground object such as a galaxy cluster will magnify and multiply images of a background object that, under other circumstances, would be not just "behind" the cluster but beyond the reach of even Webb. (If you flip back to the Webb Deep Field, you can identify many examples of these telltale arcs.) For more on the astronomy of the infant universe and Webb's contributions to that field, see Chapter Six. For more on gravitational lensing in particular, see pages 176 and 195.

PILLARS OF CREATION

PROLOGUE

The email from NASA arrives in her inbox: *Your data will be available within the month.* She waits, and eventually another message arrives: *Your data will be available within the week.* She waits, and eventually another message arrives: *Your data will arrive Sunday.*

On Sunday, however, Rebecca Larson will be flying to Seattle for the January 2023 semi-annual meeting of the American Astronomical Society.

Maybe the data will arrive before her plane takes off?

No such luck.

Maybe while she's in the air?

No.

Maybe during her layover in Denver?

Yes! But: It's raw data. Gigabytes and gigabytes of data, most of which is extraneous for her purposes. Still, it's a start, a lode Larson can begin to mine on the next leg. She and a longtime astronomer friend who's also going to the meeting, Taylor Hutchison, huddle with their laptops—but the airport's internet can't handle the load. And besides, what they really want is the processed data—the lode minus the runoff.

Maybe it will arrive by the time they land in Seattle? No. Maybe by the time they check in to their off-site hotel? No. Maybe in the Starbucks down the block, where they're killing time before the opening reception?

Yes.

The two of them text Dan Coe, the principal investigator of their team. He's seen the email too, and he suggests they round up the other members of the team and meet in the lobby of the official hotel, where Coe is staying. If nothing else, that location might have better internet access.

The several members of the team claim a set of chairs around a coffee table in the lobby. They had targeted a galaxy dating back to maybe 400 million years after the Big Bang in an attempt to demonstrate that the new telescope's instruments are capable of detecting the lines corresponding to emissions—the chemical composition of a galaxy—even at that distance across space and time.

Coe delivers the bad news. A team member in Copenhagen has already applied his own code to the data, and there's nothing there. Just "noise," a fog of static—perhaps radiation from another light source overwhelming the light from their own target galaxy.

Groans all around. Sighs. Some misting in the eyes. Everyone slumps. Coe turns his laptop so everyone can see their colleague's graph for themselves.

Larson and Hutchison lean close. They're the resident experts in identifying this kind of data. They've mastered the technique of extracting emission lines from background noise. They've been extracting data from noise for years.

"Oh, yeah," Larson says. "There's lines in this."

Hutchison agrees. "That," she says, pointing to a spot on the screen, "that right there is a line."

Coe wants to know: *Are you sure?*

He'll be able to see for himself, Larson says, once she runs the processed data through her own code.

But that'll have to wait. The lobby has filled with astronomers filing toward the opening reception across the street, and Coe and his team join the procession. But Coe can't wait. Leaning close to Larson, Coe wants to know: *Do you really think you see emission lines?*

Yes.

Over drinks: *Do you really think you see emission lines?*

Yes.

During dinner: ditto and ditto.

After dinner, at a bar with a bunch of other astronomers, Larson casually solves a colleague's coding problem, in the process producing a photo that her colleague can use in a press release, and then everybody passes that laptop around the bar, praising Larson's work. But Larson has coding problems of her own to try to solve, so she heads back to her hotel, goes to her room, and climbs into bed.

She hasn't slept in eighteen hours. She turns out the light and closes her eyes.

And then, sighing, she sits up, opens her laptop, and gets to work.

But hey, nobody ever said seeing back to the beginning of space and time was going to be easy.

"The history of astronomy," the American astronomer Edwin Hubble wrote in 1936, "is a history of receding horizons."

Which is, metaphorically, how the history of science works. One generation inherits a horizon, then tries to figure out a way to cross it. But that history is itself a combination of two tales.

One is a tale of curiosity. The other is a tale of tools. This call-and-response between vision and mission — between intellectual ambition and technological innovation, between the knowledge we want and the means we have for retrieving it — isn't unique to astronomy. It happens when one generation puts eye to microscope and wonders what an even finer view might reveal, or when one generation fires up a particle accelerator and wonders what an even more energetic collider might discover.

But in astronomy the horizons aren't just metaphorical.

Ever since Galileo pointed a primitive telescope at the night sky in 1609, astronomers have encountered one fresh horizon after another. Galileo discovered moons around another planet. Then other astronomers discovered more moons around other planets. Then two more planets. Then moons around those planets, too. They discovered stars otherwise invisible, stretching as far as their increasingly powerful telescopes could see, and they found cloudy patches that even the most powerful telescopes at the turn of the twentieth century couldn't resolve. Edwin Hubble himself, in the 1920s, determined that those cloudy patches are "island universes"—galaxies like our own Milky Way. And then in the 1990s his namesake telescope discovered that the universe was dense with galaxies as far as its eye could see—and, presumably, farther. Farther across the universe and—since light takes time to reach our eyes—farther into the past.

How much farther in space? How much farther in time?

Even before the launch of the Hubble Space Telescope in 1990, its successor—the James Webb Space Telescope (JWST)—was in the works. At that point in the history of astronomy, nobody knew precisely what horizons the Hubble telescope would identify. But astronomers were confident that it would identify *some* new horizon—a horizon that would excite the curiosity of the next generation of astronomers, the horizon for which they would need a more powerful tool in order to cross.

The original name of the James Webb Space Telescope even invoked that pattern: the Next Generation Space Telescope. Even so, it almost *missed* the next generation. Billions of dollars over budget, more than half a decade behind schedule, the telescope incurred the wrath of Congress and, in 2011, briefly went out of business.

In the end, Congress granted it a stay of execution, and on December 25, 2021—after another decade of further and further delays and greater and greater budgetary overruns—it launched. Over the next few weeks the telescope performed hundreds of technological feats, the failure of any one of which would have scrubbed the mission. But once the telescope reached its permanent home a million miles from Earth, it got to work.

In February 2022 it began funneling data back to the Webb command center in Baltimore, where engineers and astronomers would oversee the necessary technical adjustments before the telescope could start performing science. But even then, in those early weeks, the first downloads of data told them all they really needed to know.

The binary zeros and ones; then the conversion of zeros and ones via algorithms; then the generation of cresting waves on graphs measuring the motions and metallicity of celestial objects, and the generation of images of those moons, planets, stars, galaxies, spanning the universe from the *near* and *now* to the ultimate *there* and *then*:

It's going to work, they told one another, and they started laughing in relief and disbelief. *It's going to work better than we ever imagined.*

The public got its first look at Webb images during an unveiling at the White House on July 11, 2022. We saw stars in our galaxy arising out of a mass of gas and dust; five galaxies in a gravitational dance; a "deep field" yielding tens of thousands of galaxies at a glance.

But perhaps no image in the early months of Webb operations captured the public imagination more than the "Pillars of Creation." The image was an update of the iconic "Pillars of Creation" photograph from the Hubble Space Telescope in 1995: two towers of gas and dust lying at the heart of a nebulous cloud of stars some 41 trillion miles from Earth; twin ziggurats of star birth and planet formation processes that will keep those clouds churning for eons and eons to come.

Other wonders followed. Not just images but discoveries, horizon after horizon after horizon. In our solar system: water in weird places. In other stellar systems within our galaxy: the presence of potentially life-friendly elements and chemical compounds. In the hundreds of billions of other galaxies beyond our own: new ways of thinking about the growth of the universe. In the infant universe, previously inaccessible: challenges to what we thought we knew about our cosmic origins.

And then there was, if not the biggest wonder of them all, at least the wonder that made those images and discoveries possible: the telescope that is itself, if you ask the scientists who understand it best, a pillar of creation.

PART I

VISION AND MISSION

CHAPTER ONE

VISION

He had a big idea. Broad-chested, in possession of a perpetual squint, Riccardo Giacconi liked to begin his workdays making loud pronouncements about a future only he could see. Some of his big ideas caught on then and there, during an impromptu meeting in his spacious corner office overlooking the saplings on the steep embankment leading down to the Stony Run creek. Some ideas encountered resistance: A lieutenant might raise an objection, voice a caution. Sometimes the ideas that encountered resistance Giacconi himself would dismiss the following day. He was reasonable that way. Provide a powerful enough counterargument and the next morning he would beckon the lieutenant back to his office and disown the previous day's certainty.

On this occasion, his big idea was that this establishment—the Space Telescope Science Institute (STScI, or the Institute), of which he was the director, and which occupied a far corner of the main campus of Johns Hopkins University in Baltimore—should begin planning the successor to the Hubble Space Telescope.

The idea belonged to the category of those that encountered resistance, if only after a pause. First, though, Garth Illingworth, the Institute's deputy director, just stood there, blinking back at his boss. Hubble was the only thing he was working on. It was the only thing anybody at the Space Telescope Science Institute was working on. It was what the Institute was *for:* preparing for the 1990 launch of Hubble—which was still five years away.

"Oh, no," Illingworth finally said. "We don't have the time." To himself he added, *This is crazy.*

But then Illingworth reminded himself that his boss had been working on space telescopes for more than a quarter of a century—for as long as space telescopes had been a thing. For nearly as long as space *anything* had been a thing. The dawn of the Space Age—the Soviet Union's launch of the satellite Sputnik 1, the first assemblage of human design to orbit Earth and the Space Race's equivalent of the first shot at Fort Sumter—dated only to the autumn of 1957. Throughout the 1960s Giacconi had helped design above-the-atmosphere telescopes and had even led a satellite mission, one that revealed a universe awash in mysterious sources of x-rays — high-energy emissions that had shocked astronomers and now, two decades later, were still defying explanation. If anybody was an expert in the logistics of performing astronomy from space, it was Giacconi.

Illingworth decided to keep an open mind.

Giacconi then explained his reasoning. After its launch, he said, the Hubble telescope would last maybe ten years. Fifteen at the most. Even if the planning for a successor mission started right then and right there, on this morning and in this office, that next telescope wouldn't leave the launchpad for another fifteen years at the earliest. Do the math: If the next space telescope was going to follow up on whatever breakthroughs Hubble might make — and what astronomer of the Hubble generation wouldn't want the next generation to have a space telescope of its own? — then the planning would have to start right now and right here.

This morning. *This* office.

Okay, Illingworth conceded, if only to himself, *maybe this idea is not* entirely *crazy.*

For the past four hundred years, each generation of astronomers has inhabited a new universe.

Maybe that universe was new because astronomers were seeing more moons than the previous generation.

Maybe it was new because astronomers were seeing more planets.

Maybe it was new because they were seeing more stars, or maybe because they were seeing more galaxies.

Only twice, however, has a new universe emerged because *seeing* itself was new.

The first time was the autumn evening in 1609 when a professor of mathematics at the local university carried out to the garden behind his apartment in Padua an awkward instrument that had recently come to his attention. It consisted of a tube of lead and two glass disks, one at either end of the tube, and when you looked through it, it made distant objects appear near.

Galileo Galilei already knew what the magnification of distant objects on Earth could do. A few weeks earlier he'd demonstrated a slightly less powerful version of the instrument to the elders of Venice, accompanying them to the tops of towers so they could see the miraculous sights for themselves: steeples in nearby villages, seemingly close enough to touch; flags on foreign ships even before they entered the harbor. (His reward: the equivalent of tenure at the University of Padua.)

But now he wanted to know what the magnification of distant objects *not* on Earth could do. He pointed the device skyward, placed one eye at the lens near the bottom end, and, at a glance, spanned a previously unbridgeable chasm.

"We have to pursue our inquiries at a distance," Aristotle

had written in his *De Caelo,* or *On the Heavens,* almost two thousand years earlier, "a distance created by the fact that our senses enable us to perceive very few of the attributes of the heavenly bodies." Galileo's *perspicillum*—perspective tube, what future generations would call a telescope—collapsed that distance by doing what no other instrument in the history of civilization had ever done. It extended one of our five senses—and in so doing, it changed what we mean by *seeing.*

In retrospect, the astronomical definition of seeing had encompassed only what *the eye alone* could perceive. From that evening on, though, the astronomical definition of seeing would encompass what could be perceived by the eye *plus a telescope*—an instrument that relied on the manipulation of light.

In the centuries after Galileo's discovery, astronomers learned to manipulate light to better and better effect. They found that by altering the shapes of the lenses at both ends of the tube as well as the length of the tube, they could manipulate the *refraction*—the bending, and therefore the focusing—of light. Then they found that by abandoning lenses for mirrors, they could manipulate the *reflection*—or the collection, and therefore the amount—of light.

By the midpoint of the twentieth century, however, the generation of astronomers coming of age—Riccardo Giacconi's generation—were beginning to suspect that they needed to rethink their understanding of light itself.

The idea that light might reach beyond what seeing could

perceive was not new. In 1800 the German-English astronomer William Herschel replicated Isaac Newton's experiment of passing light through a prism, only this time he placed the bulbs of thermometers within the segments in the spectrum of colors, from violet through red. The thermometers, as he somewhat expected from his work on optics, registered different temperatures. The lowest were at the violet end of the spectrum and ascended toward the red. But then, beyond the red end of the spectrum, in a region that to the eye revealed no color, the temperatures kept rising. "Radiant heat," Herschel concluded, "will at least partly, if not chiefly, consist, if I may be permitted the expression, of invisible light."

THE ELECTROMAGNETIC SPECTRUM

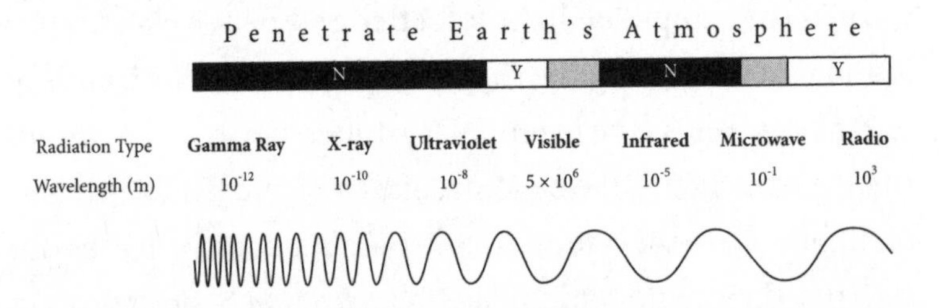

The meanings of light: Before 1800 *light* meant what we could see with our eyes, whether on their own or with the aid of a telescope. But by the midpoint of the twentieth century astronomers had begun to realize that the non-optical regions of the electromagnetic spectrum provide a wealth of other information about the universe. Webb, by seeing predominantly in the infrared, can observe farther across space—and therefore back in time—than any earlier telescope.

In the century and a half since Herschel's discovery, astronomers had learned that all light is a combination of electricity and magnetism, and that the electromagnetic spectrum spans from radio waves through microwaves through infrared light through visible light through ultraviolet light through x-rays to gamma rays. They had also learned that the speed of light is unvarying, and that what distinguishes these categories of light is the length of the wave from one crest to the next. Visible light, for instance, consists of wavelengths between 0.4 and 0.7 microns (a micron being 0.001 of a millimeter, or about 0.000039 of an inch). Moreover, within that 0.4- to 0.7-micron sliver of the electromagnetic spectrum, finer gradations in wavelengths determine the colors we perceive.

Not until the middle of the twentieth century, however, did astronomers realize that light with wavelengths outside the visible part of the spectrum—shorter than 0.4 microns and longer than 0.7 microns—might have applications for their own science. During World War II the Allies encountered strange radio signals that they attributed to German jamming. The real culprit, British engineers eventually realized, was solar flaring—an eruption of electromagnetic radiation on the surface of the Sun.

After the war, those engineers learned that in the 1930s a radio antenna at Bell Labs, in New Jersey, had serendipitously discovered that the stars in our galaxy were a source of radio waves.* One of the engineers, Bernard Lovell, persuaded the British government to finance a radio telescope with a diameter of 250 feet. In August 1950 astronomers aimed that telescope at a nearby galaxy and passed its signals through a chart recorder, a seismograph-like instrument that produced the telltale ink tracings on paper identifying the detection of radio signals. "It was no longer possible," Lovell later recalled, "to regard the local galaxy as in any way unique as a radio transmitter."

The electromagnetic spectrum is vast, but only radio and optical (and some ultraviolet) light can penetrate the Earth's atmosphere. If astronomers wanted to know whether other portions of the electromagnetic spectrum held any surprises, they had nowhere to go but up.

They sent V-2 rockets bearing Geiger counters and other detectors into the sky above the New Mexico desert, reaching altitudes as high as rockets could go: just high enough to leave the atmosphere before returning to Earth. In 1946 a detector aboard a suborbital rocket found the first evidence of ultraviolet light from the Sun. Two years later, researchers confirmed that our local star was also a source of x-rays. But only with the onset of the Space Age proper—when rockets could reach the velocity necessary to enter Earth orbit—could astronomers begin to search the universe beyond the

* Bell Labs showed no interest in pursuing the discovery. The engineer, Karl Jansky, had fulfilled the assignment to identify the source of an annoying hiss that was disrupting transatlantic radiotelephone transmissions, and that was that.

solar system for invisible light outside the optical, radio, or ultraviolet regimes. In 1962 a rocket bearing a Geiger counter located the first extrasolar sources of x-rays, including one mysterious object radiating an x-ray output ten *billion* times that of the Sun.

The lead author on the resulting paper, "Evidence for X-Rays from Sources Outside the Solar System," was Riccardo Giacconi. Soon thereafter he got to work on the first satellite that would perform x-ray astronomy exclusively: Uhuru, which launched in 1970. During its three years of operation it discovered binary x-ray sources (two sources of x-rays orbiting each other), identified a possible candidate for a black hole, and provided the data for the all-sky catalogue of x-ray emissions. Yet even as Uhuru was preparing to launch, Giacconi was already part of a team proposing the *next* x-ray satellite, Einstein, which eventually launched in 1978.

The other regimes of the electromagnetic spectrum were going through the same investigatory cycle: Send something up to see if anything is out there; refine the next mission accordingly; and even while making those refinements, design the *next* mission. By the late 1970s, NASA was developing a Great Observatories program — four satellites that, over the following two-plus decades, would explore x-rays (a mission co-proposed by Giacconi), infrared rays, gamma rays, and visible light, along with a bit of ultraviolet. The visible-plus-a-bit-of-ultraviolet instrument would be the Hubble Space Telescope, which would launch first, perhaps in the early 1980s.

For some astronomers, however, that two-decade timeline for studying the nonvisible parts of the electromagnetic spectrum was problematic; they felt that NASA had waited too long. Giacconi himself had witnessed a gap in continuity in his field. The Einstein mission, which, like the Uhuru mission, pre-dated the Great Observatories program, ended in 1981, but a follow-up x-ray mission — the one that was part of the Great Observatories initiative — now wasn't due to launch until the mid-1990s at the earliest. In 1981, facing a long lull in his career, Giacconi accepted the directorship of the nascent Space Telescope Science Institute* — where, when the time came, he might help convince NASA not to make the same mistake again.

When Illingworth, then an astronomer at Kitt Peak National Observatory in Arizona, heard that Riccardo Giacconi was accepting the position of director at the Space Telescope Science Institute, he wondered why. Giacconi was an x-ray astronomer. Hubble was going to be primarily a visible-light telescope. But once Illingworth joined the Institute as deputy director in 1984, he began to understand: What the project needed wasn't someone with a background in conducting visible-light astronomy but someone with a background in navigating big-science bureaucracy.

Even then Giacconi was using his influence to begin framing any future discussions of a successor to Hubble. In

* Not that the perk had been a deciding factor, but the director's office did come with a private bathroom.

1984, the same year that Illingworth joined the Institute, NASA commissioned the Space Science Board, an advisory agency independent of NASA, to produce a wish list of space projects for the decades ahead—specifically, from 1995 to 2015. The board divided the project into six categories (Earth Sciences, Life Sciences, and so on) and assigned one task group to each. Giacconi was a member of the task group on Astronomy and Astrophysics.

Like the other task groups, his met multiple times, beginning in the summer of 1984 and ending in January 1986. In June 1986 the Space Science Board's overall steering committee met to discuss the findings of the individual task groups. These meetings had two purposes. One was to finalize recommendations for further study. The other was to sift through the dozens and dozens of goals that the six task groups had identified and then prepare a comprehensive report. The steering committee duly sifted, sifted some more, and then, in the end, shrugged: The overwhelming majority of the projects, they concluded, were deserving of further consideration.

The resulting report, *Space Science in the Twenty-First Century: Imperatives for the Decades 1995 to 2015,* which the National Academy Press published in 1988, consisted of seven volumes—one for each of the six disciplines, plus an overview. The preface to the overview conceded that the report offered no road map, no hierarchical flowchart. It offered, instead, a smorgasbord: an assortment of delights in no particular order. The steering group didn't even bother to rank its recommendations. Nonetheless, among the recommendations in the "Astronomy and Astrophysics" chapter in the overview was "an 8- to 16-meter Telescope" that would "follow on 10 to 20 years of study with the HST" and "provide images 6 times sharper than the [2.4-meter] HST."

Yet even if the point of the report wasn't to assign priorities to projects, it did have some advice for NASA: Reassign *your* priorities.

On January 28, 1986, the space shuttle *Challenger,* shortly after leaving the Cape Canaveral launchpad on the east coast of Florida, having reached an altitude of 46,000 feet, disintegrated. An examination of the relatively intact crew compartment a few weeks later, after its recovery from the floor of the Atlantic Ocean, revealed that the emergency-oxygen intake of at least some of the seven astronauts exactly matched the two-minute-plus duration of the descent, beginning with the disintegration of the rest of the shuttle and ending with the crew compartment's impact with the Atlantic. They'd been alive.

For many members of the U.S. space-science community, the "accident" (as NASA called it, though the more common description was "disaster") was vindication, however bitter. They had long questioned NASA's treatment of space science—its allocations of funding and choices of research programs. Their objection wasn't that NASA was ignoring their field. Instead, it was that almost from the moment President Dwight D. Eisenhower signed into law the National Aeronautics and Space Act of 1958—and certainly ever since May 25, 1961, when President John F. Kennedy told a joint session of Congress, "I believe this nation should commit itself to achieving the goal, before this decade is out, of landing a man on the moon and returning him safely to the Earth"—space scientists had found themselves competing with a *concept*, a Cold War imperative masquerading as a romantic ideal: what the overview to *Space Science in the Twenty-First Century* summarized as "humans in space."

"For the past 30 years," the overview continued, explicitly invoking the 1958–1988 extent of NASA's lifetime to that point, "scientific investigation has been neither the only objective of the space program of the United States, nor even the dominant one. The Apollo project and the development of the Space Transportation System"—space shuttles—"and, more recently, of the Space Station were not primarily designed to respond to requirements set by the various disciplines of space science." The report didn't exactly state that the continuity of astronomical investigation—the centuries-long tradition of generation-after-generation inheritances of new universes and the inventions of new means for investigating them—had lapsed at the literal expense of the space shuttle program. But it didn't need to: "The steering group for this study recommends that the present ordering of priorities in the national space program be changed."

Having made its point, the report then immediately—in the very next sentence—made its point again, only now in italics: "*The steering group proposes that, as the nation considers its future in space, the advance of science and its applications to human welfare be adopted and implemented as an objective no less central to the space program of the United States than any other, such as the capability of expanding man's presence in space."*

Then the report made its point yet again, not just later in the same introduction but later in the same paragraph: "This will ensure that the scientific and engineering resources available are effectively utilized in the national interest, as required by the act of 1958. This same standard—obtaining the greatest scientific advance for the available resources—should prevail when determining the balance between manned and unmanned space activities as well."

Anyone at NASA reading this introduction might have paused at this juncture, then flipped back a couple of pages to reread the essay's epigraph to see if it had been saying what they now suspected it might have been saying. If so, they would have found a quote from Proverbs 29:18—one that might have seemed, a few minutes earlier, like inspirational pablum appropriate for a multivolume publication of the National Academy Press, but which now had acquired an acidic, *Challenger*-specific subtext:

Where there is no vision, the people perish.

So what would be the Space Telescope Science Institute's vision for the successor to Hubble?

That question, implicit in Giacconi's challenge to Illingworth, was actually two questions in one—the same two that drive scientific advances generation after generation.

First: What does the community want to investigate?

Second: What technology would allow it to make those investigations?

But answering those two questions was hardly a straightforward process. Instead, it came with a complication, a level of complexity that was endemic to the scientific method: An answer to one depended on an answer to the other.

Science, contrary to common perception, rarely proceeds in a straight line from hypothesis to experiment to validation (or refutation). At any moment scientists might possess a proposition to test without having the means to do so. Or they might possess the means to test a proposition, if only they had one to test. Because of that interdependence, the scientific process is piecemeal—a herky-jerky, start-and-stop progression that might or might not reflect progress. Scientists have a term of art for this kind of narrative: *nonlinear.*

Illingworth's assignment was, in effect, to make the progression toward a successor to Hubble as linear as possible. Capitalize on the mention in *Space Science in the Twenty-First Century.* Pluck a Hubble successor from the smorgasbord of future missions and recast it as a fait accompli. Grab it off the groaning board and plate it.

Which Illingworth promptly did. In August 1988, soon after the publication of the *Space Science* report, he delivered a talk at the International Astronomical Union meeting in Baltimore. "The first step has been taken," he said, citing the report's mention of a post-Hubble project, "and we need to move on." Continuity, he insisted, was essential. Without mentioning Giacconi by name, Illingworth invoked his boss's experience: "I am sure that any astronomer, particularly the observers, can imagine the impact on their research programs if access to their primary facility was cut off for ten to fifteen years. It would be devastating, yet this is what has happened in X-ray astronomy. The Einstein satellite had a dramatic impact in many areas. It offered a tantalizing glimpse of data on many important problems—and then the door was shut."

If the community could keep the door to a Hubble successor open, what would they want the telescope to do? One obvious answer was to see what Hubble could see, only better—to observe objects at greater distances and at a higher resolution. Another obvious answer was to see what Hubble *couldn't* see—and already two such targets were emerging. They were in the *Space Science* report, and they were in Illingworth's speech at the International Astronomical Union convention. A successor to Hubble should target galaxies not long after the birth of the universe, and planets around other stars within our own galaxy. Observing either target would require pushing past the visible part of the electromagnetic spectrum on the longer-wavelength end: the infrared, which picks up where the visible regime ends, at a wavelength of about* 0.7 microns. The infrared regime then continues into the hundreds of microns, but even a telescope that could detect at the 20-plus micron level would be a major advance.

At a level of 20-plus microns, a telescope in principle would be able to see galaxies as far away in space (and, the speed of light being finite, as far back in time) as a couple of hundred million years after the Big Bang, maybe even earlier. In an expanding universe such as ours, what's expanding is space itself, and the expansion of space in turn stretches the light from galaxies as it journeys across the universe. By the time the light from galaxies in the first billion years of the universe reaches us, the expansion of space has shifted its wavelengths beyond the visible regime of the electromagnetic spectrum and into the infrared—a phenomenon that astronomers call *redshift*.

Seeing in the infrared would also allow us to pierce the dust that surrounds star-forming regions in our own galaxy. Other infrared space observatories had gotten tantalizing glimpses, but astronomers wanted to blow through the "fog" and see those stellar nurseries in detail.

By imagining such an instrument, as Illingworth knew, astronomers were placing their faith in the continuing advances of current technology. But, he had to wonder, why would technology stop advancing?

---

* *About* because a spectrum is, after all, a spectrum.

His generation was old enough to remember using computer punch cards—flimsy rectangles of heavy paper that a programmer had to manually poke holes in before loading a stack of them onto a feeder tray on the side of a console the size of a Volkswagen. Now, though, astronomers were using the-future-is-here Unix workstations, complete with color screens, standalone keyboards, and floppy disks.

And telescopes themselves were still making a fundamental transition in technology from photographic plates (which astronomers, working in the dark, had to lick to make sure the side with the emulsion was facing the right way in the camera) to charge-coupled devices (CCDs). Whereas a photographic plate could soak up maybe five percent of the available photons, a CCD detector could collect upward of eighty percent.

These and other technological advances were transforming astronomy, but would they continue at such a pace that a successor to Hubble would make enough of a scientific difference to justify the time and money?

Leaps of faith, Illingworth thought. A succession of *if* and *if* and *if.*

If the technology is there. *If* Hubble finds galaxies distant enough to justify a search for earlier ones. *If* extrasolar planets even exist.

But most important of all: *If* the project doesn't go horribly, ruinously, mission-scrubbingly nonlinear.

Go nonlinear it did, even before Illingworth could execute the next step in his effort to rally the community into wanting to make a successor to Hubble a reality.

By 1989 Illingworth had left the Space Telescope Science Institute to be an astronomer at the Lick Observatory, part of the University of California Observatories, but he continued to work closely with NASA and the Institute. He and Pierre-Yves Bely, an engineer at the Institute specializing in telescope design, scheduled a workshop, "The Next Generation Space Telescope," at the Institute for mid-September 1989. By summer preparations were well underway: topics chosen, talks scheduled, speakers invited. But on July 20, the twentieth anniversary of Neil Armstrong's first steps on the surface of the Moon, President George H. W. Bush stood before the National Air and Space Museum and declared the nation's intent to return. "In 1961 it took a crisis — the space race — to speed things up," he said. "Today we don't have a crisis; we have an opportunity. To seize this opportunity, I'm not proposing a ten-year plan like Apollo; I'm proposing a long-range, continuing commitment." Which was, he specified: "Back to the Moon; back to the future. And this time, back to stay."

The Moon, the Institute liaison at NASA informed Illingworth, was now part of the "Next Generation Space Telescope" workshop — or it would be if the Institute wanted NASA's buy-in. And so the topic of a telescope on the Moon joined that of a telescope in space: new sessions, new speakers...and little enthusiasm. If putting a telescope on the Moon proved to be the only way to get a successor to Hubble, then fine: You take what you can get. In the meantime, though, the consensus in the community was to keep focusing on the more likely eventual outcome of a telescope in space.

Still, Illingworth had a message he needed to impart to the community in order to keep the project linear. In the editors' introduction (Illingworth was one of three) to the published proceedings of the workshop, the second paragraph echoed Giacconi's call to action from a few years earlier: "It might seem too early to start planning for a successor to HST. In fact, we are late."

Only eight months later the project encountered another nonlinearity, though this time far more serious, even potentially fatal.

The space shuttle ferrying Hubble launched on April 24, 1990. On May 20 Hubble saw first light — the telescopy term for the initial observation. Over the following month the telescope radioed back images that carried an unmistakable message: *Baltimore, we have a problem.*

The primary mirror was out of focus.

An astronomical mirror needs to be slightly curved so that no matter where on the reflecting surface light lands, it will bounce back to a single focal point — the secondary mirror (where it will then bounce to a detector, whether eyeball or CCD chips). The edges of Hubble's mirror, though, were off by 1.3 millimeters, more than enough to cause a significant distortion.

The mission immediately became a national punchline. Late-night comics and water-cooler wits mocked the fiasco as a classic example of *You had one job. . . .* The error rendered Hubble unable to perform many of its duties, leaving NASA vulnerable to congressional oversight bordering on strangulation. At the Space Telescope Science Institute, work on a space telescope for the next generation all but vanished. First they had to fix *this* generation's.

But what if they couldn't?

Hubble continued to perform some breakthrough science — observing the ring-of-gas remnants of a supernova, for instance, and imaging a disk of material as it disappeared into a possible black hole. But back at the Institute, the all-hands-on-deck, around-the-clock focus (so to speak) was on how to fix the mirror. By the summer of 1993, the Institute had figured out how to do so, and NASA had ceded time on a space shuttle mission that December during which space-walking astronauts would now make the repairs.

But that commitment came with a caveat that few people knew about, even within the upper echelons of astronomy.

That August, Alan Dressler, an astronomer at the Carnegie Institution of Washington, received a phone call from Goetz Oertel, the director of AURA — the Association of Universities for Research in Astronomy, a cost- and facility-sharing consortium of research centers dating to 1957 but also the government-independent administrator of the Carnegie Institution.

In a sense, Hubble was just one more telescope under AURA's jurisdiction. But it was obviously more than that: Hubble was the telescope with the greatest government investment and the highest public profile. Yet Hubble was also less than that: NASA had always been running the show, whatever a bureaucratic flowchart might claim to reflect. And NASA, the director of AURA told Dressler, was ready to abort the repair mission.

The human risk was high. Everybody knew that. Strapping people onto projectiles and blasting them hundreds of miles off terra firma was always a risk. But now NASA, weighing costs and benefits, was thinking the risk might be too high. NASA, Oertel told Dressler, had made clear that they were not going to unduly risk anybody's existence — or, with it, NASA's. They had therefore informed AURA that they were not only reserving the right to cancel the mission but to do so in consultation with absolutely nobody else. And, Oertel added, NASA was placing the odds of cancellation at fifty-fifty.

Fascinating, Dressler thought. But what, he wondered, did AURA want with him?

"We have nothing on the drawing boards," Oertel said. If Hubble was no longer a going concern, AURA would still be in charge of telescopes around the world. But it wouldn't be in the generation-defining space-telescope business. Would Dressler be willing to chair a committee to discuss that contingency? To come up with an alternative to Hubble, just in case?

In other words, the eternal questions:

First: What does the community want to investigate?

Second: What technology would allow it to make those investigations?

That autumn, Dressler convened a committee of twenty astronomers, including himself. They talked about what kind of telescope the Hubble replacement should be. Like Hubble? Not like Hubble? Looking for exoplanets? Looking for early galaxies?

A successor to Hubble, of course, *had* been under discussion for as long as any space scientist of the Hubble telescope era had recognized that a *new* instrument was nearly always a precursor to a *next* instrument. It had already been under discussion, however vaguely, even before Riccardo Giacconi joined the Space Science Board's committee advising on the future of space science. And while Giacconi's instruction to Illingworth, two or three years later, to start planning for whatever was *next* had taken his deputy by surprise, the directive had soon made sense. A successor to Hubble was inevitable, as long as someone could will it into existence.

But the ultimate result of all those discussions — the ideas that emerged from the Institute workshop in 1989 — hadn't impressed Dressler.

Infrared, fine. Infrared, *of course.*

You want to cut through the dust obscuring the visible-light view of our galaxy's stars and their planetary systems? Then *yes,* infrared.

You want to probe the infant era of the universe, before the expansion of space stretched the light emanating from the first visible objects beyond the optical-light portion of the electromagnetic spectrum? Then *yes,* infrared.

But the instrument in the 1989 report would have merely included infrared in the mix, along with ultraviolet and visible wavelengths. All that report had proposed, in Dressler's opinion, was *Hubble, only bigger.*

Worse, it had proposed a *Hubble, only bigger* model before Hubble had even launched—before anybody knew what Hubble could do. That approach, Dressler thought, was "stillborn." He felt that even if Hubble hadn't been out of focus, the 1989 report was still a classic case of cart/horse. In 1991 the Astronomy and Astrophysics Decadal Survey, the National Research Council's once-every-ten-years-or-so ranking of funding requests for future projects, hadn't even mentioned a proposal that Illingworth and his colleagues had submitted.

Dressler's HST [Hubble Space Telescope] & Beyond Committee convened while NASA was still deciding whether to proceed with the December 1993 repair mission. In the end, NASA's risk/reward analysis gave the mission the go-ahead. And that mission was a success. Hubble was now ready to try to perform the kind of cutting-edge science that NASA and AURA had feared it would need to abandon. An immediate alternative to Hubble — the alternative that Oertel had asked Dressler to spitball — was no longer necessary. Dressler, however, saw no reason to disband the committee.

Now — now that the world was about to find out what Hubble could actually do — *now* was the time to decide on a successor to Hubble.

What Hubble could do turned out to be a surprise for astronomers in two ways.

One was scientific.

The telescope was back in business, and business was booming. In May 1994 Hubble validated the existence of supermassive black holes — the sources of those x-ray emissions that Giacconi and a Geiger counter had discovered in 1962.* That July the telescope relayed to a waiting world images of a comet's fragments as they plowed through the clouds of Jupiter, creating craters of atmospheric disturbance, each of which dwarfed the size of Earth. In February 1995 Hubble discovered oxygen on Europa, one of Jupiter's many moons. The revelations were arriving with such frequency that astronomers could hardly believe their good fortune at

* Giacconi would share the 2002 Nobel Prize in Physics for his "pioneering contributions to astrophysics, which have led to the discovery of cosmic X-ray sources."

being astronomers at this particular moment in the history of science.

The other surprise was maybe not so much a surprise as a shock. It was a shock for Dressler, anyway. And from his conversations with other astronomers as well as the members of the administration at NASA, he sensed that it was a shock for them, too.

The public loved Hubble.

Loved it. Couldn't get enough of it. Color photos began appearing everywhere—on the front pages of newspapers, across magazine spreads, in books, on television programs, and, increasingly, via the World Wide Web.

For NASA veterans of a certain age the cultural moment was redolent with nostalgia for the 1960s. For the first time since the heyday of the Apollo program the public was experiencing the exploration of space as a human endeavor—though with one significant difference.

Back then, the exploration of space happened through surrogates. It meant identifying with the person reclining in a capsule or planting a boot on the Moon. It meant standing in your backyard and looking at the heavens and knowing one of your own was up there.

Now, the exploration of space meant seeing the universe for yourself. Seeing it up close and in color. Seeing it—*really* seeing it—for the first time. You could sit at your desktop computer (not just a desktop computer at work, but a desktop computer *in your own home*) and plant the mouthpiece and earpiece ends of a landline phone into the corresponding cushions of a modem, dial up the provider number, wait through a soft-jazz version of an MRI exam, monitor the horizontal bar at the bottom of the screen as it slowly accumulated a measure of whatever percentage of the way the website had loaded, and then, if you were lucky, see an image that might alter your understanding of space and time.

In December 1995, Dressler visited the Washington, DC, office of Daniel S. Goldin, the administrator of NASA. Goldin had accepted the directorship of NASA three years earlier, after twenty-five years at TRW, the mega-corporation (more than a hundred thousand employees) where he had overseen the execution of some of the Great Observatories missions and eventually had ascended to the directorship of space programs relating to intelligence gathering and the military. Goldin's response to Dressler's suggestion that NASA begin planning for a successor to Hubble wasn't surprising: "For God's sake, it's now working for the first time." Besides, the rescue mission had demonstrated that Hubble's life span wasn't the fifteen years ending in 2005 but...longer.

Goldin, Dressler knew, wasn't an astronomer. He was primarily what his NASA title implied, an administrator. But, Dressler assumed, Dan Goldin was also human, and humans like stories. The story Dressler now told Goldin was maybe the grandest of them all.

Dressler was, by his own admission, "sentimental" about science. He could trace this trait to several formative moments in his life, but the one that came to mind now was a lecture he'd attended at Caltech maybe fifteen years earlier. For an astronomer at the nearby Carnegie offices in Pasadena, these colloquia were like going to the movies, and sometimes the speaker was indeed the scientific version of a marquee attraction: a Richard Feynman, a Murray Gell-Mann. On this occasion, the lecturer was Luis Alvarez, a Nobel laureate in physics, a colleague of J. Robert Oppenheimer's on the Manhattan Project at Los Alamos,* and, coincidentally, a former business partner of one of Dressler's uncles.

Alvarez, however, wasn't there to talk about high-energy physics. Instead, as Dressler realized minutes into the lecture, Alvarez wanted to talk about some findings his geologist son Walter had made and a theory the two of them were developing. As far as Walter Alvarez could tell, his father Luis said, sediment all around the world contained a layer rich in iridium dating to around 66 million years ago—but iridium isn't common in the Earth's crust. It is, however, common in asteroids. Could an asteroid's collision with the Earth, Alvarez *père* and *fils* wondered, have caused the extinction of the dinosaurs?

Dressler couldn't get the lecture out of his head. For weeks he thought about it. In his scientific experience, the disappearance of the dinosaurs had always seemed to be one

* Like going to the movies indeed. Alvarez was a minor character in the 2023 film *Oppenheimer.*

of those mysteries we'd never solve. But what if the Alvarezes were right?

Even if they weren't right, what if they *could* be right? An asteroid that destroyed the dinosaurs would have cleared the landscape for the emergence of our own species. What if the discovery of those post-apocalyptic first stirrings was possible? What if—Dressler thought, a couple of decades later, as he considered the images emanating from the Hubble Space Telescope—we could trace not just the emergence of our own species but the universe as we know it?

Now Dressler placed on the table before Goldin two items. One was *The Hubble Atlas of Galaxies,* a collection of photographs of nearby galaxies by Edwin Hubble's protégé Allan Sandage. Dressler opened it to an image of the Andromeda Galaxy, a spiral galaxy like our own Milky Way. Then Dressler gestured to the other item he'd placed on the table, a recent image from the Hubble telescope showing a lot of dots of light against a background of a lot of dark.

Galaxies, Dressler said. He pointed to this dot, and this dot, and this dot. Each of these dots, he said, is a galaxy. And each of these galaxies, he went on, is a galaxy like our own. And each of these galaxies contains, like our own, billions of stars. And those stars are the hosts of who knows how many planets?

But Dressler set those questions aside for the moment.

These galaxies are telling us a *story,* Dressler said. The farther we can see, the deeper we can peer into the past; the deeper we can peer into the past, the closer we can come to the beginning of that story. We can learn a lot about that story with Hubble, but that telescope can see only so far. If we could see farther, Dressler said, we might be able to see the start of the story. The *Once upon a time* part. The beginning-of-the-universe part. Not in the Big Bang sense, but in the sense of the first galaxies, the first stars.

In the sense of—Dressler said, invoking a word that he had been hearing in his head lately, over and over, the word that had become indistinguishable from the story he wanted the telescope to tell—our *origins*.

Also in attendance at that meeting were the head of the NASA Space Science Division, NASA's chief scientist, and, most important, Ed Weiler, NASA's director of astrophysics and therefore NASA's liaison with the Space Telescope Science Institute, AURA, and all things Hubble. Once Dressler had finished his presentation, Weiler suggested that Dressler step outside and wait in the hallway. After about ten minutes, Weiler emerged from Goldin's office.

Great news, he said. Dressler's sales pitch had worked. Goldin was ready to green-light a successor to Hubble.

The following month, on January 15, 1996, in a presentation at the American Astronomical Society meeting in San Antonio, NASA released a composite image that it was calling the Hubble Deep Field.

From December 18 to December 28, Hubble had stared into a single pinprick of the sky. Pick your scale of comparison: the diameter of a tennis ball at one hundred meters, maybe, or a dime at seventy-five feet, or a grain of sand at arm's length. Astronomers had chosen this particular pinprick because it was relatively free of foreground stars in our galaxy. What they wanted to see was what lay *beyond* our galaxy.

More galaxies, sure.

But how many? And how distant?

For eleven days, over 150 earth orbits and 342 exposures, the photons from whatever was out there flooded the telescope's CCD detectors, then flooded them some more, and then kept flooding them, drilling through the equivalent of a core sample of the universe, a hole of seeming darkness that covers 1/24,000,000th of the entire sky. By the time Hubble turned its attention elsewhere, it had registered at least two thousand galaxies at a distance of up to 12 billion light-years, or well more than three-quarters of the way back to the Big Bang.

Hubble had reached its limit, and in doing so, it made Dressler's point—only this time not to one NASA administrator in Washington, however influential in determining the future of the space program, but to the world. Squinting at the Deep Field's faintest whorls, you couldn't help wondering what else was out there.

How far could we see?

How far across space? How far back in time?

In effect, the Hubble Deep Field defined a new horizon, one that endowed a successor to the Hubble Space Telescope—a space telescope for the next generation—with a vision to rival humans in space: the first light in the universe.

CHAPTER TWO

MISSION

Two days after NASA unveiled the Hubble Deep Field image at the January 1996 semiannual meeting of the American Astronomical Society, the director of NASA, Dan Goldin, gave a major speech: "NASA in the Next Millennium." The January meeting always draws several thousand astronomers, and now most of them had crowded into the ballroom in San Antonio to hear about the future — of astronomy, of course, but also of their careers.

At one point in his talk, Goldin turned his discussion of space science in the next millennium to the prospect of a successor to Hubble. He paused for a moment, then narrowed his attention to an attendee in the front row.

"I see Alan Dressler here," he said.

Dressler straightened in his seat. That the head of NASA would mention him by name within the context of a talk about the future of space science in the next millennium wasn't exactly a surprise. Dressler was the chair of the HST & Beyond Committee. He would be speaking next, after Goldin, summarizing for the community the recommendations of the HST & Beyond Committee to date. Still, Dressler had no idea what Goldin was about to say.

"All he wants," Goldin went on, looking down at Dressler, "is a four-meter optic"—a mirror with a diameter of four meters. "And I said to him," Goldin continued, " 'Why do you ask for such a modest thing? Why not go after six or seven meters?' "

The dozens of rows of astronomers behind Dressler roared.

In their meeting in Goldin's office the previous month, Goldin had indeed suggested to Dressler that a successor to Hubble might house a mirror much larger than the one on any proverbial drawing board. And Dressler's first thought had been *That's fantastic!* But by the time of the AAS meeting, he'd had second thoughts—the same second thoughts he knew everyone else in the auditorium would soon be having, once they heard the catch: the same catch that he'd heard in Goldin's office.

Which they then heard: The budget wasn't budging. The diameter of the mirror would double, but the cost of the project—half a billion dollars—would remain the same.

No roars now. Just murmurs.

"Mike, it's going to be a thermal nightmare."

Frank Martin, the head of astrophysics at Lockheed Martin, was offering Mike Menzel the job of chief systems engineer on a proposal for the Next Generation Space Telescope. Two or three years had passed since Goldin's rousing, then puzzling, announcement at the January 1996 American Astronomical Society meeting. In the interim NASA had solicited proposals from the giants of defense and aerospace, and three had responded: TRW (a company later acquired by Northrop Grumman), Goddard Space Flight Center, and Lockheed Martin.

As the deputy program manager at Lockheed Martin for Hubble's servicing missions, Menzel was already familiar with the technology of space telescopes. Now he took one look at the spreadsheet listing the telescope's technical requirements and stated the obvious.

"Yes," he said, "it's going to be a thermal nightmare."

An infrared telescope in space comes with multiple technical challenges. Since infrared radiation is sensitive to heat, the detector would need to be cold — the nearer to absolute zero (–273.15 degrees Celsius or –459.67 degrees Fahrenheit), the better. It would therefore have to be both beyond the reach of Earth's warming effects and impervious to direct exposure to the Sun.

But everybody working on the project already knew that.

An onboard coolant system was out of the question, because it would deplete fuel and shorten the life span of the mission. The instrument would therefore have to cool down on its own—a process that instrumentalists call passive cooling—by allowing space itself to act as a coolant.

But everybody working on the project already knew that, too.

And the inclusion of a shield to block the heat of the Sun in order to maintain those low temperatures: also common knowledge.

Put all these restrictions together and you do indeed get a "thermal nightmare." Menzel, however, had identified another challenge, one that was subtler than lowering the temperature of an object in space and keeping it low.

"It's also going to be a thermal nightmare to *verify,*" he said.

His about-to-be boss gave him a look.

"It's not *testable,*" Menzel said.

The problem was the sunshield. In his work on Hubble, Menzel had become familiar with NASA's institutional culture. NASA, he knew, liked to perform its verifications "on the ground"—in an environment that duplicates as closely as possible the conditions in space under which instrumentation will eventually be operating.

Aerospace campuses house chambers that can approximate numerous extreme physical conditions. At the Goddard Space Flight Center in Greenbelt, Maryland, a multistory chamber contains speakers capable of producing sounds at a decibel level that, if a human were inside the chamber, would gelatinize eardrums. In preparing for a NASA mission, engineers could avail themselves of facilities expansive enough to contain an instrument that is beyond massive. They could also use facilities capable of reproducing beyond-extreme physical conditions such as the ultraviolence of a rocket launch or the longueurs of deep space. They could even use facilities capable of testing a beyond-massive instrument undergoing beyond-extreme physical conditions.

But how do you test a telescope that will be operating at temperatures in the hundreds of degrees above zero, whether Celsius or Fahrenheit, on one side of a sunshield while also operating at temperatures in the hundreds of degrees below zero, whether Celsius or Fahrenheit, on the other side of the sunshield? How do you create an artificial environment in a testing facility the size of an aircraft hangar in which the temperature must vary by hundreds and hundreds of degrees—whether Celsius or Fahrenheit—over the course of a few meters?

You don't.

"You're going to have to prove this thing is flightworthy by analysis," Menzel told Martin.

Analysis was different from performance. Analysis required physical testing, too, but ultimately it relied on math. Analysis meant coming up with a mathematical model for how each half of the project would fare under its own set of extreme conditions, then separately testing each half operating under that set of extreme conditions. If the mathematical model for one half of the project matched the results in one test chamber, and the mathematical model for the other half of the project matched the results in another test chamber, then you would physically join the two halves of the project together and trust that the mathematical model for the whole would match the results in the ultimate test chamber: space.

Martin—whose long career included working on systems operations for Apollo 16 and Apollo 17—suggested that Menzel write up these insights. In 1998 Menzel published an article, "A Strawman Verification Program for the Next Generation Space Telescope," that essentially established him as the Solomon of aerospace, only better: He could sever the two halves of the telescope's operations without sacrificing the baby.

In the end, Northrop Grumman won the competition. But Lockheed Martin didn't lose just the NASA contract. It lost Mike Menzel, whom Northrop Grumman hired at NASA's behest. Then, in 2004, Northrop Grumman also lost Menzel—to NASA, which offered him the position of lead systems engineer on the Next Generation Space Telescope. Menzel had to take a fifteen percent pay cut to move from the private sector to the public, but he and his wife, Catherine, ultimately decided the financial sacrifice was worth not just the professional advancement but the personal challenge.

He would be joining the defining astronomical project of his generation. Menzel would be on a mission. He and the thousands of other scientists enlisting in the project wouldn't quite be reinventing the wheel. But they *would* be reinventing the telescope.

To say that the following two decades were an exercise in scientific nonlinearity would be an understatement of literally cosmic proportions.

Many of the nonlinearities dated to the moment Dan Goldin peered down from a podium in San Antonio, singled out Alan Dressler in the front row, and implicitly endorsed a space telescope with a diameter of six or seven meters at the same $500 million cost as a space telescope with a diameter of four meters. Which is not to suggest that $500 million would have covered the cost of a space telescope with a diameter of four meters, either, as seemingly everyone else in space astronomy already knew.

At the 1989 workshop at the Space Telescope Science Institute, one astronomer had estimated the cost of a ten-meter space telescope at $3.8 billion (about $9.5 billion in 2024). The following year, Illingworth, in a paper he submitted as part of a presentation to the 1990 Astronomy and Astrophysics Decadal Survey, estimated a budget of $2 billion. Even though the *HST and Beyond* report that Alan Dressler's committee published in mid-1996 continued to press for a four-meter telescope rather than Goldin's six- or seven-meter alternative,* the budget estimate of $500 million was no longer the target but the base. Maybe, the report said, the cost would reach $1 billion.

Goldin arrived at NASA having made a name for himself in the aerospace industry for his frugality. He was a proponent of an approach to major projects that he encapsulated in a catchphrase: "Faster, better, cheaper." He had a reputation for delivering projects on time (or earlier) and at (or under) budget, and once he joined NASA and tried to apply that philosophy to its institutional culture, he concluded that NASA's aversion to risk was itself a risky approach. His reasoning: By consolidating disparate objectives into as few flights as possible, NASA had inadvertently encouraged scientists to lard their missions with more instruments, leading to longer waits and greater weights, both of which resulted in higher costs for the spacecraft and the launcher.

Still, consensus in the space-exploration community was that if the head of NASA was offering to bankroll a project that could redefine your field, you said yes and hoped that more money would come your way once the unrealistic nature of the original budget became obvious even to the head of NASA. Indeed: By 1998 the official budget for the Next Generation Space Telescope had increased to $1 billion, the upper end of Dressler's range.

---

* "Hubble huggers," Goldin dismissively called the four-meter advocates.

Even so, attaching a psychological sense of inevitability to the mission — the campaign that Giacconi, Illingworth, and others at the Institute began in 1985 and which the Hubble Deep Field reenergized in 1996 — would help too. In 1997 NASA appointed an Ad Hoc Science Working Group to begin providing advice about vision and mission. In 1999 Ed Weiler, NASA's director of astrophysics, officially designated the Next Generation Space Telescope as a NASA priority, and in 2001 the all-important Astronomy and Astrophysics Decadal Survey emphatically reinforced that status: The Next Generation Space Telescope was NASA's new top priority.

That sense of inevitability prompted NASA's new director, Sean O'Keefe, the former secretary of the navy and deputy director of the Office of Management and Budget who had replaced Goldin, to perform the ritual renaming of a telescope-in-progress. No longer was Hubble's successor to be the Next Generation Space Telescope. Now it was the James Webb Space Telescope.

The Next Generation community paused in its preparations. The practice of replacing generic names for telescopes and observatories with the names of prominent scientists was routine — Galileo, Cassini, Spitzer, Chandra, Hubble, and on and on. O'Keefe, however, had violated two norms: His choice of honoree was a unilateral rather than a consensus-of-the-community decision, and that honoree was not a scientist but an administrator — one of O'Keefe's predecessors,

James E. Webb, who had served as the director of NASA from 1961 to 1968, during its Space Race heyday.

And then the Next Generation — now Webb — community resumed its preparations. In 2006 *Space Science Reviews* devoted an issue to Webb's objectives and design. The lead paper — by the members of the project's Science Working Group, which was a successor to both the Ad Hoc Science Working Group and an Interim Science Working Group — finalized the mission's four goals, which in turn reflected the trajectory of the narrative that Alan Dressler had advanced on a desktop in Dan Goldin's office in December 1995. Webb would be a journey across the universe, a tale in four acts:

- Planetary Systems and the Origins of Life;
- The Birth of Stars and Protoplanetary Systems;
- The Assembly of Galaxies;
- The End of the Dark Ages: First Light and Reionization.

Before long the James Webb Space Telescope acquired another aura of inevitability. Yes, it was going to happen — but it was going to do so on its own increasingly intractable terms: perpetually behind schedule and over budget.

By 2000 the cumulative cost of its predecessor, the Hubble telescope, had amounted to $4 billion, adjusted for inflation. But because of the "faster, better, cheaper" approach, NASA even then was still estimating that Hubble's successor would cost $3 billion *less* than Hubble—that the Next Generation Space Telescope could meet the $1 billion price tag that the 2001 Decadal Survey had specified. And because Webb's launch dates and budget had never been realistic, every year NASA had to convey to ("lie to" was Dressler's characterization) the U.S. Office of Management and Budget what fresh infusion of funding would keep the project operational.

Under the circumstances, unrealistic projections of time and money were almost inevitable. Webb engineers were inventing technologies at a level of complexity and innovation nobody had ever attempted—a 25-square-meter (275-square-foot) mirror, for example, that could fold up or unfold, or a sunshield consisting of five layers, each layer the length of a long tennis lob and the width of a tissue.

And Webb production managers did find ways to ameliorate costs. In 2007 NASA finalized agreements with the Canadian Space Agency and the European Space Agency that had been under discussion since the late 1990s: In exchange for a share of observing time on the telescope, the two agencies would develop and deliver key Webb instruments. In addition, the European Space Agency would provide the launch site in French Guiana as well as an Ariane 5 rocket to ferry the telescope into space.

Even so, a fantasist approach of underestimating the launch dates and budget became routine during that formative decade, merely a part of the process. At one point NASA approved instrument upgrades but didn't adjust the 2010 launch date, as if designing and implementing the upgrades would take no time at all—literally.

As Webb devoured more and more resources, leaders of non-Webb projects at NASA as well as researchers in the community at large found their own hearts' desires suffering delays (and therefore their own careers suffering setbacks). By 2010 the Webb budget had risen to $5 billion and the launch date had slipped to 2014. The 2010 Decadal Survey selected the Wide-Field Infrared Survey Telescope as the top priority, but the leaders of that project tempered their celebrations when they learned that their project's development would have to wait until the top priority from the *previous* decadal survey had launched—whenever that might be. A headline in a 2010 issue of *Nature* characterized the status quo as self-cannibalization: THE TELESCOPE THAT ATE ASTRONOMY.

On June 29 of that year, Senator Barbara Mikulski, Democrat from Maryland (home of both the Space Telescope Science Institute in Baltimore and the NASA Goddard Space Flight Center in Greenbelt), in her role as chair of the subcommittee overseeing NASA's budget, sent a letter to NASA. "I am deeply troubled by the escalating costs for the JWST," Mikulski wrote. "Simply put, NASA must manage the cost and schedule of its large-scale programs to the highest standard."

Her "request" was that NASA "initiate an independent and comprehensive review of JWST." Oh, and make it snappy: Get to work within thirty days, as "this panel's input will be critical to our consideration of NASA's FY 2011 appropriations."

Charles Bolden, a NASA administrator as well as a former astronaut who had flown on four space shuttle missions, piloting two (including the one that deployed Hubble) and serving as mission commander on the other two, responded with a three-paragraph letter agreeing in full. The concision itself was a sign of obeisance: *Say no more.* The panel even took its name directly from Mikulski's request for "an independent and comprehensive review": the Independent Comprehensive Review Panel (ICRP).

Garth Illingworth wound up being the only astronomer (and the only scientist, for that matter) to serve on the panel. He assumed he got the invitation thanks to his familiarity with large NASA projects, with the science community and its objectives, and — dating back at least to NASA's stipulation that the 1989 Institute workshop include discussions of a telescope on the Moon — with the politics of big science. In recent years he had chaired the National Science Foundation's Astronomy and Astrophysics Advisory Committee, and in that capacity he had consulted with the NASA administration at its highest levels, met with politicians and their staffs on Capitol Hill, and, once, appeared before a congressional committee to discuss the nation's astronomy program.

The review panel submitted its Independent Comprehensive Review Report to Congress in November 2010. Part of the behind-schedule and over-budget mindset, it concluded, arose from longstanding mismanagement: "The institutional cost and program analysis capability at the Directorate and Agency level continues to function without the skill and authority required. Consequently," the report continued, "demand for corrective action was absent, and poor management practices on the JWST Project went unquestioned for too long." The Webb administrators' nearly pathological habit of lowballing budget requests, knowing that they could always go back for more funding in the following cycle, had finally caught up with them.

The Independent Panel elected to reject fantasy and embrace realism. Even at this late stage in the review process, it said, NASA was still being overly optimistic. Citing NASA's current estimates of "a 70% probability of launching in June 2014 at a total lifecycle cost of nearly $5 billion," the committee's report rhetorically asked if the project could possibly meet that schedule or budget. The conclusion: "No chance." Instead, the report projected an "earliest launch date" of September 2015 and a "minimum cost" of $6.5 billion.

Mikulski had chosen the November 2010 deadline for a reason. On November 2, voters would decide whether the Congress that would convene two months later was to remain Democratic or become Republican. The verdict: Republicans won in a landslide. Worse, from the perspective of Webb leadership, not just Republicans but Tea Party Republicans—the fiscally conservative kind who were demanding steep reductions in federal spending. That month the House, still under the control of Democrats for another six weeks, kept Webb in the budget, leaving NASA to hope for the best with the next Congress — a process that would begin by correcting the problems cited in the Independent Review report.

The first problem — mismanagement — NASA addressed by removing the project manager at the Goddard Space Flight Center. Illingworth, for one, thought the manager had done a good job, under the circumstances. But Illingworth also recognized the need for political expediency: *Somebody's head has to roll.* Other heads followed: the astrophysics director, the program executive. (Partly as a result of this bloodbath, the ICRP gained an alternate acronym within the Webb community: ICRAP.)

Addressing the second problem — cost and delays — was trickier. The report insisted that the new leadership would need to make their own assessments, since the ICRP members had been working on a tight time frame and, despite Illingworth's presence as the resident expert, couldn't possibly understand all the nuances of the mission. The new Webb administrators consulted with their staff and came back with new estimates: a launch date of not 2015 but 2018 and a cost of not $6.5 billion but $8.8 billion.

At least one member of Congress had decided: *Enough is enough.* In July 2011 Representative Frank Wolf, Republican of Virginia and chair of the U.S. House Appropriations Committee on Commerce, Justice, and Science, called for the cancellation of the project.

Many in the Webb community thought Wolf must be bluffing. Others weren't so sure. History, after all, was not on their side. You couldn't tell yourself that Congress wouldn't *really* cancel a scientific mission of such ambition (and sunk costs), because once, it had. In October 1993 President Bill Clinton signed a bill killing the Superconducting Super Collider, which would have been the world's most powerful particle accelerator. Never mind that the project had already swallowed $2 billion. Never mind that underground boring in the open prairie near Waxahachie, Texas, had already cleared nearly nineteen of the eventual fifty-one miles of tunnel. Never mind that the particle accelerator promised transformative scientific breakthroughs. Congress deemed the project's budget to be out of control. The cancellation had blown a hole through the heart of the U.S. particle-physics community, which had no choice but to cede the Superconducting Super Collider's potentially world-beating scientific stature to the more diminutive (27-kilometer, or 17-mile, circumference) Large Hadron Collider, then under construction on the French-Swiss border.

In November 2011, Congress delivered its verdict on the fate of Webb. Yes, they would continue to fund it, but they would do so with a caveat, an inviolable budget cap:

Eight billion bucks or bust.

"If they're going to cancel, that's fine," Mike Menzel would tell his crew over the coming years. "Don't worry. Don't even listen."

On a fairly regular basis, Menzel would have to appear before a review board—exercises that his longtime Hubble telescope confrère and Webb collaborator Peter Stockman once likened to Kabuki theater—to explain why the project was taking so long or costing so much. His standard answer was pretty much what he said when Frank Martin first showed him the plans for the observatory at Lockheed Martin back in the late 1990s:

Relying on analysis—completing a mathematical model for the part of the satellite on one side of the sunshield, then completing another mathematical model for the part on the other side of the sunshield, and then seeing if the models might match—contained a greater capacity for human error than putting the whole kit and caboodle in a several-story-high chamber and shaking it as if there were no tomorrow (which, in the event of stress-test failures, there might not be, in terms of the mission's future).

What he needed, Menzel would tell the review boards, was "margin"—shorthand not just for *margin of error* but for *margin of error beyond the margin of error.* "The normal rules don't apply here," Menzel liked to say. "This is virgin territory"—a landscape of foreseeable challenges but, more important for his purposes, also unforeseeable threats: "unknown unknowns," as the Webb crew came to call the problems they couldn't imagine.*

"How much margin do you need?" one program manager or another would periodically ask Menzel.

"As much as I can get," he'd answer.

Budget overruns, bureaucratic ineptitude, congressional oversight, review-board reckonings, the whole process of rethinking how to test a space telescope from the ground up: Webb had survived them all. One other factor, however, continued to create havoc with the budget and the launch timeline deep into the 2010s—what Menzel called "stupid mistakes."

One such mistake was miswirings that had fried some of the prototypes' electrical components—for instance, the pressure transducer for the propellant (the transducer being, more or less, the gas gauge). *Do we fly without pressure transducers?* Menzel's team had to debate the question. The verdict: *No.* So they had to replace them.

---

* The phrase was a deliberate echo of the U.S. defense secretary Donald Rumsfeld's explanation of the difficulty in gathering intelligence during the Iraq war. "Now what is the message there?" he said on June 6, 2002. "The message is that there are known knowns; there are things we know that we know. There are known unknowns; that is to say there are things that we now know we don't know. But there are also unknown unknowns—the ones we don't know we don't know." The snickering started then, in that briefing room at NATO headquarters in Brussels, and it hadn't stopped since. But Rumsfeld was right. Unknown unknowns are a standard consideration, not just in warfare but in science.

Another stupid mistake: the application of an inappropriate solvent that damaged the observatory's propulsion valves.

And another: seven tears in the sunshield.

Another: a vibration test for the sunshield that ended with dozens of bolts blowing loose and bouncing around the test chamber. The problem turned out to be that the bolts had too few threads. (Team members wound up fishing bolts out of far crannies of the facility for months afterward.)

Partly because of these accidents, the launch date slipped from October 2018 to June 2019. The U.S. Government Accountability Office, after conducting an investigation into the delay, released an analysis cautioning that even a launch date in June 2019 was likely too optimistic, and indeed a month after the Accountability Office had issued that analysis, NASA announced a further delay, to spring 2020. It also acknowledged that Webb had reached Congress's $8-billion-or-bust budget cap... and would need to exceed it, if the telescope was ever to get off the ground.

In January 2019 Congress approved a further infusion of $800 million, bringing the total expenditures to $8.8 billion. An accompanying report was brutal. "There is profound disappointment with both NASA and its contractors regarding mismanagement, complete lack of careful oversight, and overall poor basic workmanship on JWST," the report said. "NASA and its commercial partners seem to believe that congressional funding for this project and other development efforts is an entitlement, unaffected by failures to stay on schedule or within budget." And once again Congress threatened the existence of the project: "NASA should strictly adhere to this cap or, under this agreement, JWST will have to find cost savings or cancel the mission."

"Just forget about them," Menzel would tell his team. "I don't care what they say. If you see a problem, just say so, and if we've got to delay, we'll delay." He compared the final preparations for launch to the folding of a parachute: "One little mistake and we're dead meat."

Then came Covid, and with it a work slowdown, prompting the announcement, in July 2020, that the telescope would launch no sooner than October 31, 2021.

For two years engineers had been assembling the components of Webb at NASA's Jet Propulsion Laboratory, just outside Pasadena. Now the time had come for the telescope to begin its journey from Long Beach to the launchpad at the European spaceport off the northeast coast of South America, near Kourou, French Guiana. The spin of the Earth adds momentum to a rocket launch, and that spin is greatest at the equator; Kourou is only three hundred or so miles north of it. But the telescope couldn't be shipped as if it were simply one more piece of cargo aboard a freighter. It required a special temperature- and humidity-controlled container. (The freighter itself had the opposite requirement: It needed to be as nonspecial, as anonymous, as possible, so as to guard against the possibility, however remote, that pirates might seize the ship and hold its $8.8 billion cargo hostage.)

The observatory survived the sixteen-day, 5,800-mile passage down the west coast of Mexico, through the Panama Canal, up the Kourou River to the Port de Pariacabo, and into a processing facility near the launch site in French Guiana without incident — but then it suffered a jolt. A high-tension clamp band snapped off, shaking the observatory. An inspection revealed no damage, but the late-December launch slipped another few days.

Even the name of the telescope, in those final weeks, became a distraction, albeit of the public-relations kind. The tenure of the historical personage of James Webb as the second in command at the Department of State in the late 1940s and early 1950s and then as the head of NASA in the 1960s coincided with what historians had come to call the "lavender scare" — a search for and purge of homosexual employees at federal institutions (ostensibly because they might be subject to blackmail and therefore pose a security risk). Investigations had turned up scant specific evidence of James Webb's involvement, but the association between bureaucracy and bigotry was close enough that some astronomers decided they would sacrifice linguistic convenience and thereafter refer to the telescope not as "Webb," a one-syllable nibble, but only as "JWST," a six-syllable Dagwood sandwich.*

Despite the mission's history of nonlinearity over the previous decades, it couldn't have reached this final stage

* Webb, not JWST, is the name that has become the norm among nonscientists. This book reflects the public consensus.

without a lot going right. You could miss that perspective if you were in the eye of the maelstrom or even closely observing it. Even so, no one involved in the project could know *how* much had gone right until the Ariane 5 rocket bearing Webb had not just lifted off but survived "six months of terror" (as the media liked to say). During those months the observatory would confront 344 "single points of failure," in NASA nomenclature—executions of technology that would reveal whether the miracles of invention and ingenuity that were unique to this mission actually worked on-site (that is, in space), the failure of any one of which would scuttle the entire project.

Those first six months, though, weren't what the scientists at the Institute and NASA were most concerned about. It was, instead, the first thirty days. Or maybe even the first two weeks, the period during which Webb would perform some of its most intricate exercises.

A forecast of high winds for December 24, 2021, forced the launch to slip one last time. On the morning of December 25, in the second-story command center at the Institute, in the auditorium downstairs on the main floor, at viewing parties at aerospace facilities around the world, and in front of computer screens on every continent, humans gathered to watch and wait.

At 7:20 a.m. Baltimore time—Baltimore time thenceforth being the official timestamp for the receipt of Webb's nonterrestrial communications—the European Space Agency's countdown to the launch reached that relic from the 1960s Space Race: the backward count to blastoff.

Dix…
Neuf…
Huit…
Sept…
Six…
Cinq…
Quatre…
Trois…
Deux…
Unité…
…*
Décollage!

Those "two weeks of terror" were not the most excruciating two weeks in Mike Menzel's life. He'd been there before.

The previous time, the terror had been unrelenting. Day after day had brought no relief. But at the one-week mark, a squib of good news emerged: His nine-year-old son's chemo had started to kick in. That fact alone, though, offered only a vague possibility of hope. It certainly didn't portend optimism. And so more days of uncertainty passed: a period of

* [*Pause dramatique.*]

helplessness, an eon of doing nothing but waiting. Not until the end of the second week could Menzel begin to relax, because only then did the doctors decree that his son's cancer was in remission.

Menzel couldn't help notice the unsettling parallel between the two time frames. "This is the second most nerve-racking two weeks of my life," he told close friends as the holiday season of 2021 bled into 2022. Still, that earlier personal experience helped put even the worst-case scenarios of the telescope's commissioning into perspective — and perhaps helped explain his sanguine attitude toward the ongoing possibility that Congress would shut down a mission that had occupied half of his professional life. He'd seen worse.

He'd seen better, too — the remission of his son's leukemia and the restoration of his health. But even if the first two weeks of Webb's life in space couldn't match Menzel's personal experience, they were turning out to be pretty spectacular.

The launch itself had been flawless. The telescope's separation from the launch vehicle, half an hour after liftoff, had also been flawless. But those tasks, while potentially catastrophic, were routine, at least within the patently absurd context of rockets blasting off into space and payloads separating from launch vehicles. The first real test of the ingenuity of Webb's engineering would come three minutes after the separation of telescope from launch vehicle: the deployment of a solar array that would provide battery-saving energy.

The Mission Operations Center (MOC), on the second floor of the Space Telescope Science Institute, consisted of two rooms. In the front room, fifty or so controllers sitting at computer monitors were relaying instructions to the observatory and watching a live feed from the rocket. In the back room, on the other side of a window, were the engineers, including Menzel. He had already informed his team that at the moment of the solar-array deployment the relevant mechanisms would be facing away from the Sun.

"So we're not going to see it?" somebody looking over his shoulder said.

Menzel shook his head.

"Mike!"—a muffled cry.

Menzel turned around. John Durning, the deputy project manager and someone Menzel considered not just a close colleague but a friend, was calling to him through the glass. Durning gestured to a monitor carrying the live feed.

"Look at this!" Durning called.

What? Why? The array wouldn't be deploying for another minute or two.

Menzel looked, and what he saw was pure saturation. A light overexposure that rendered the image meaningless.

The observatory was pointing at the Sun.

"Holy shit," said Menzel.

Menzel had based the estimate of when the solar array would emerge on the assumption that three minutes was how long the observatory would be rocking from the aftershock—what space engineers call the "tip-off rate" (*tip-off* meaning separation; *rate* meaning the resulting rate of rotation). Only when the telescope had sufficiently steadied would the on-board computer deploy the solar array. The saturation, however, meant that the deployment was earlier than anyone had expected, suggesting that the separation of telescope from rocket was smoother than anyone had expected.

Only thirty minutes after launch, the telescope had already banked a bunch of fuel. *More margin!*

Twelve hours after launch, right on time, came the first trajectory correction. Two days later came a second. Eight days after that, when the telescope was 500,000 miles from Earth, or halfway to its destination, came the deployment of the sunshield—or at least the beginning of the deployment.

The sunshield fell in the category of unique-to-this-mission. The Sun-facing side of the observatory would have a temperature of 110 degrees Celsius (230 degrees Fahrenheit), while the rest-of-the-universe side would need to be about –237 degrees Celsius (–394 degrees Fahrenheit), a difference of more than 600 degrees Fahrenheit. Webb technicians had created an aluminum- and silicon-coated design consisting of five layers. The first layer, facing the Sun, was 0.05 millimeters (0.002 inches) thick. The other four layers were twice as thick as the first, though "thick" was relative: Instead of one-sixtieth the width of a human hair, these layers were one-thirtieth. Altogether the distance from the first layer to the last was four meters, and cumulatively they provided an SPF — Sun Protection Factor, as in sunscreen lotions — of 1 million. Upon unfolding, the first layer would measure twenty-one by fourteen meters (sixty-nine by forty-six feet); the other layers would be smaller, but only nominally so.

The unfolding process commenced on December 31. It would involve releasing a total of 107 mechanisms, several at a time. The first ninety-three releases worked, according to the sensors monitoring the motions. The release of the final bank of fourteen mechanisms, however, had not.

In the back room of the Mission Operations Center, Menzel and his team stared at the monitors. They were pros. On the outside, they maintained their cool. On the inside, though, they knew: *This is the nightmare scenario.* If the last of the release mechanisms had in fact failed, then the sunshield was going to rip, and when it did, the telescope, one hundred million person-hours, three-plus decades, nearly $9 billion, and very likely the fate of NASA would follow it into the abyss.

Minutes passed.

Nothing.

Half an hour passed.

Nothing — nothing to report other than a bunch of engineers standing around, being professional, trying not to fidget so as not to trigger panic in their peers.

But then, about an hour after the deployment (or non-) of the sunshield, someone piped up behind Menzel.

"Wait a minute." It was a thermal engineer. He was looking at the output from a thermistor—a thermal sensor—that was aboard the spacecraft in the vicinity of the deployment mechanisms that they thought might have failed. At the precise moment that mission control had fired the 107th mechanism, he said, that nearby thermistor had gone from hot to cold. Which suggested, he explained, that something had blocked the thermistor from the Sun. What could the source of that shadow be other than the possibly errant mechanism—a roll-up cover that would indeed be in the precise position to block the Sun, assuming that it had deployed?

While nobody had seen the deployment, the team could infer that it had happened. A potentially mission-ending crisis had turned out to be merely a pulse-racing pause.

The next few months went like that, mostly.

On January 4, 2022, the observatory deployed the telescope's secondary mirror. "We are 600,000 miles from Earth," the Webb project manager Bill Ochs announced to his team in the Mission Operations Center in Baltimore, "and we have a telescope."

Not quite. Everyone within earshot knew that in reflecting telescopes—the kind with mirrors—incoming light hits the primary mirror at the base of the telescope first, then bounces back to a much smaller secondary mirror, which redirects the light sideways toward the instruments. And everyone knew what Ochs meant by "We have a telescope." But everyone also knew they wouldn't *really* have a telescope until not just the secondary mirror but the primary mirror deployed.

On January 6 and 7, the primary mirror—or, more accurately, eighteen hexagonal mirrors that together acted as one giant mirror—did just that. This kind of design had been an integral part of astronomy only since the early 1990s, when the single-mirror, or monolithic, lens had reached its technological limit. The larger the mirror, the greater the light collection; but the larger the mirror, the more difficult and more expensive the grinding process, and the more massive and more expensive the supporting structure. At a certain diameter—from six to eight meters, or twenty to twenty-six feet—mirrors had exhausted the technology necessary to make them, but even if they hadn't, the cost would have been prohibitive. Enter the segmented mirror, a mosaic of smaller hexagonal mirrors in the shape of a honeycomb. The first telescopes to adopt the segmented mirror innovation at a significant scale were, in 1993 and 1996, Keck I and Keck II at the Mauna Kea Observatories, atop a dormant volcano on the Big Island of Hawai'i. Segmented mirrors had since then become standard for the most ambitious observatories, a category that included Webb.

Webb's primary mirror would technically be under the eight-meter-diameter outer limit, but a monolithic lens of that size would be too heavy to launch. Then again, so would eighteen smaller mirrors, at least if they followed the standard glass-mirror approach. So the Webb engineers (partly at the suggestion of Goldin) chose instead to make the mirrors out of the relatively lightweight but strong element beryllium, with a coating of gold.

Yet the weight of the mirrors was not the most difficult design challenge. It was their size. The collective diameter of the primary mirror would span more than 6.5 meters, or 21.6 feet (in contrast to the 2.4-meter, or 7.9-foot, diameter of Hubble's mirror), far too wide for a rocket's fairing—the nose cone housing the payload. So engineers developed an ingenious solution. They divided the honeycomb into sections that would fold up and therefore fit inside the rocket on Earth. Then, in space, the sections would unfold origami-like.

On January 24, following one last course correction, Webb reached its final resting place (so to speak): a region of space that astronomers call the second Lagrange point. L2 is one of five sites in the solar system that the nineteenth-century Italian-French mathematician Joseph-Louis Lagrange determined would keep pace with Earth in their orbits around the Sun. At a Lagrange point, the gravitational balance between Earth and the Sun acts as a stabilizing influence. The orbit around the Sun of an object at a Lagrange point would be in perpetual sync with the orbit of Earth, as if the L2 passenger were hitching a ride.

As with other spacecraft,* this gravitational convergence offered a practical advantage for Webb. By riding shotgun

---

* Among the astronomy projects that have occupied L2—actually, still occupy L2, though they're no longer operational—are the Wilkinson

with Earth, Webb would be expending minimal fuel (other than shifting in its seat from time to time in order to get a better bead on a horizon), thereby conserving fuel and significantly prolonging the life of the mission.

For an infrared observatory in particular, L2 has a further advantage: It's always in Earth's shadow. It exists in a state of perpetual solar eclipse. L2's position on the side of Earth directly opposite the Sun reduces exposure not only to light but also to heat that would hopelessly compromise observations in the infrared.

A few days after Webb reached L2, mission control began activating the telescope's four instruments.

For the photons that would soon be sacrificing themselves for science, the eighteen mirrors were only the first stop. After bouncing off those surfaces, they converged at the secondary mirror, where they ricocheted to one of Webb's four scientific instruments. Three of those instruments would be covering the same wavelength range, from 0.6 to 5 microns, in a complementary fashion.

The Near-Infrared Camera, the product of a collaboration between the University of Arizona and Lockheed Martin, would be the telescope's primary imager in the near infrared. It came with a coronagraph — an opaque disk that serves as the astronomical equivalent of your hand when you want to block the glare of the Sun. The feature is especially

---

Microwave Anisotropy Probe and the Herschel and Planck space observatories.

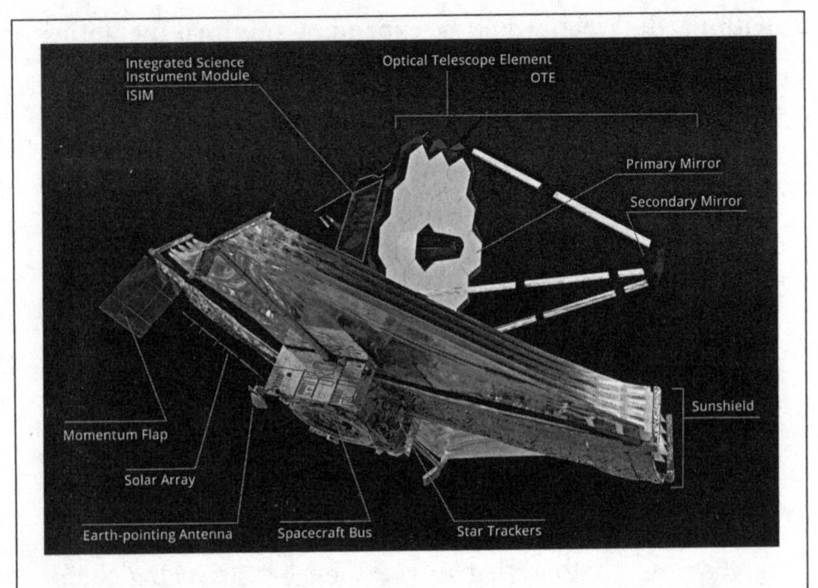

Kit, meet caboodle: Webb consists of two major parts — the side facing toward the Sun, and the side facing away from the Sun. Separating those two halves is the sunshield. Because the two parts experience temperatures hundreds of degrees apart, whether Celsius or Fahrenheit, engineers couldn't test them together in the same chamber. Instead, the "test chamber" was space itself. On the Sun-facing side are the solar array and the spacecraft bus, the home to Webb's support functions (Electrical Power Subsystem, Attitude Control Subsystem, Communication Subsystem, Command and Data Handling Subsystem, Propulsion Subsystem, Thermal Control Subsystem). On the other side of the sunshield is the telescope. Photons first hit the eighteen segments of the primary mirror, then bounce to the secondary mirror, which redirects them into the Integrated Science Instrument Module, where Webb's four astronomical instruments (the Near-Infrared Camera, the Near-Infrared Spectrograph, the Mid-Infrared Instrument, and the Fine Guidance Sensor/Near-Infrared Imager and Slitless Spectrograph) process them according to the specifications that astronomers have predetermined will meet their data-collection needs.

useful in the study of planets around stars. By blocking the bright light of a star, the coronagraph would allow the instrument to detect the light of orbiting planets and debris.

The Near-Infrared Spectrograph, the product of a collaboration between the European Space Agency and Airbus Industries, would have the usual capacity to gather data that reveals a target's physical and chemical properties. But, via NASA, it also housed a microshutter array of 248,000 doors that controllers could manipulate individually so as to gather spectra from up to one hundred objects or points in space at once.

The Near-Infrared Imager and Slitless Spectrograph, the product of a collaboration between the Canadian Space Agency and Honeywell International, included a masking option that, in effect, would transform Webb into an interferometer—an instrument that sends light from a single source down different optical paths. In the case of Webb, the separation of light would allow it to capture images of bright objects at a resolution greater than that of the other imagers.

The fourth instrument, however, operated almost entirely out of the wavelength range of the other three. The Mid-Infrared Instrument, the product of a collaboration between the European Consortium and the Jet Propulsion Laboratory, covered the 4.9- to 27.9-micron range—deep into the infrared portion of the electromagnetic spectrum, where the expansion of space would have stretched the once-visible light of the most distant, and thus the earliest, objects in the universe. The instrument would therefore need to be more sensitive to heat contamination than the other three Webb instruments (since heat produces thermal radiation that shows up on the infrared spectrum). Unlike the other three instruments, it couldn't cool down sufficiently simply by sitting in space, waiting for the

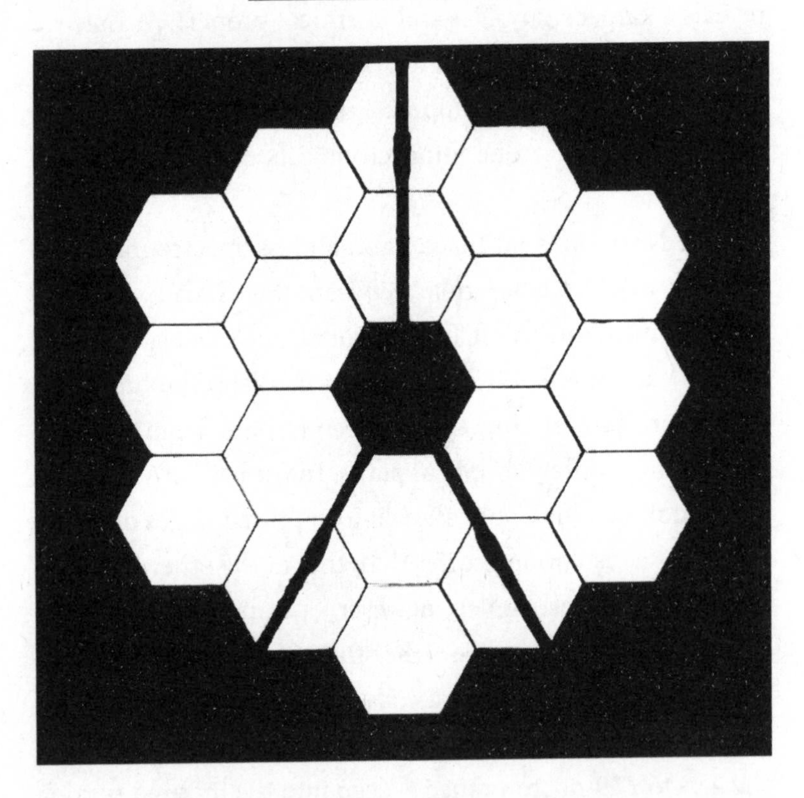

A Webb selfie: A special, non-astronomical camera aboard Webb captured this view of the eighteen mirror segments that together make up the primary mirror. The camera was there specifically to help engineers on Earth align the individual segments so that they produce one uniform image.

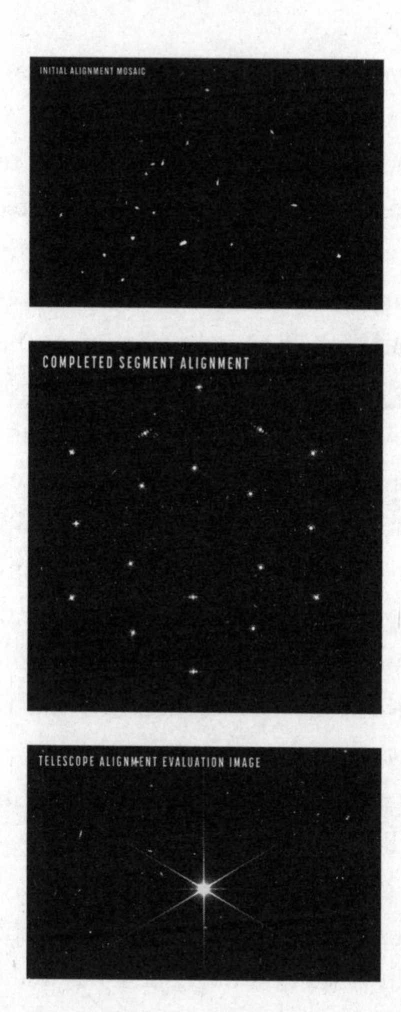

[Top] At "first light," the eighteen mirror segments reflected eighteen images of the same star. Note the bright spots in the background: "Those aren't stars," Mike Menzel said to a colleague when they first saw this image. "The big ones — they're stars. All the rest are *galaxies*." [Middle] After engineers executed numerous adjustments to the individual mirrors, the eighteen images matched. [Bottom] The composite of those eighteen images — officially, the "Alignment Evaluation Image."

cosmic vacuum to do its thing. Instead, it carried a helium refrigerator. That cryocooler system, however, would need to labor for another month before the instrument reached the target temperature of 7 degrees Celsius (13 degrees Fahrenheit) above absolute zero — dozens of degrees cooler than its siblings.

On February 2 the telescope opened its eyes — a moment that astronomers call "first light." Engineers had preselected a single star as the test target. Each of the telescope's mirrors reflected its own image of the star, but no two mirrors revealed the same image. As everyone expected, the location of the star within the images varied from mirror to mirror. Then, operating under the direction of the optical telescope element manager, Lee Feinberg, actuators on the back of the primary mirror adjusted the individual segments until the multiple images of the star aligned. The star was still out of focus, but it was *there:* one star.

"Mike," Feinberg said to Menzel, "this thing is going to work beautifully."

The two engineers had gathered in Menzel's office to review the first composite images. Menzel agreed: Webb was going to work beautifully. Then he thought he noticed something.

He leaned close to the computer screen.

In even the most single-star-centric views of our Milky Way galaxy, you would expect to see a few stars in the background. After all, our galaxy contains multitudes. It hosts billions of stars. Point an ultrapowerful space telescope at any peephole in the heavens and you'll almost always see some dots of light other than the one you wanted to study. Menzel pointed to a few of them.

"Lee," he said, "those aren't stars. The big ones—they're stars. All the rest are *galaxies*." He started laughing. "Holy shit," he said. "Look at them all. And we haven't even focused this sucker yet."

Nineteen hours, more than fifteen hundred test images, and fifty-four gigabytes of data later, they focused the sucker.

Webb was going—dare they dream it?—linear.

Over the following months technicians continued to monitor the progress of the mission around the clock. In deference to Covid, they adopted mandatory on-site masking and social distancing, and—at least as important—they devised ways of also working from home, all without sacrificing science. At the Institute, they could avail themselves of food trucks that parked in front of the building, along San Martin Drive, some offering standard food-truck meals, others offering more exotic fare such as crabmeat sandwiches, a Baltimore delicacy. They adorned their desktops with stress-reliever squeeze balls as well as good-luck charms that, however unscientific,* seemed to do the job.

Occasionally the on-board computers would sense a glitch and send the observatory into "safe mode," essentially shutting down the telescope. The engineers at the Institute would spend an hour or two figuring out what the problem

* Very.

was and creating a software patch, and then all systems would be good to go again.

And so passed the "six months of terror," percolating along without a major event—with one scary exception. Scary to some, anyway. Sometime in late May—nobody could say when for sure—a micrometeoroid slammed into one of the eighteen mirror segments that made up the primary mirror.

Everyone knew that this would happen at some point. Space is hardly empty. It's vast, certainly. But our solar system is still a work in progress. It's been under construction for only four and a half billion years, and although the primordial accretion of gas and other detritus into a central star, planets, and moons is long past, the solar system is still aswirl with cosmic sawdust. The observatory's engineers had even rehearsed the impacts of micrometeoroid-mass objects on mirror samples, since such events were inevitable.

So the fact that on a random day in May 2022 and in a random patch of L2 the trajectory of a micrometeoroid through time and space intersected with an object was no surprise. The telescope had already registered four such impacts. What *was* a surprise, however, was the size of the micrometeoroid. It was on the high end of the mathematical calculations for what the primary mirror could endure.

Was the size of the micrometeoroid a fluke? If so, no problem.

But what if it wasn't a fluke? What if it was a harbinger? A first example of a regular occurrence? If so, the telescope would still be able to absorb a rain of similarly sizable rocks, though as a result the mission would probably barely survive its minimal projection of five years.

“Are you worried?” a project manager said to Menzel.

Menzel laughed.

“For twenty years, you’ve been accusing me of hoarding margin,” he said. “I’ve got margin to spare, and *this* is why I can sleep.”

On July 12, 2022, Webb’s vision and mission officially converged. On that day, Webb made the transition from test mode to science mode. From then on, the observations would belong to the astronomers who had won time on the telescope—who had described what they wanted to observe, why they wanted to observe it, and how Webb would be indispensable in helping them achieve those goals, one vision at a time.

When Edwin Hubble characterized astronomy as “a history of receding horizons,” he could have been describing the distinct realms that would define Webb’s goals.

The first horizon: the boundary between our world and the planets and moons in the solar system.

The second horizon: the stars and planets beyond the solar system, filling the rest of our Milky Way galaxy.

The third horizon, the one that Edwin Hubble himself had crossed: all the galaxies beyond our own.

The fourth and final horizon, in terms of Webb's goals, was one that Hubble couldn't have anticipated, although the fact that a horizon beyond the galaxies awaited future generations of astronomers wouldn't have surprised him: the first light in the universe.

The day before Webb made the transition from test mode to science mode, President Joe Biden unveiled the first public image at a ceremony in the Eisenhower Executive Office Building, directly across the street from the Oval Office. The image that would introduce Webb's virtuosic powers of perception to the world had remained a secret except among a few Webb insiders, including Menzel. It was an update to the image that, more than a quarter of a century earlier, had carried such a visual impact that without it Webb might never have reached the drawing board, let alone the launchpad: the Hubble Deep Field.

"It looks like the Hubble Deep Field," Catherine Menzel said to her husband when she, along with virtually the rest of the world, got her first look at Webb's Deep Field during a live video feed from the Executive Office media event.

Mike Menzel needed a moment to gather himself.

"Yeah, you're right, honey, it does. But Hubble took fourteen days to take *that* picture. We did *this* in twelve hours, and in twelve hours the faintest things broke Hubble's records. And we weren't even *trying*."

That was what he said to his wife.

What he'd said to his colleagues, though, when he first saw the image, was "Holy shit."

PART II

FOUR HORIZONS

CHAPTER THREE

FIRST HORIZON: CLOSE TO HOME

Heidi Hammel was at a loss for words.

Which was uncharacteristic for her. She had long before distinguished herself as a consummate public advocate for astronomy, an expert who could make complex ideas compelling for nonscientists. She often received requests for help in communicating astronomy to the public. Sometimes a request came in the form of an invitation to give a public presentation — a TED talk or a lecture at a university. Sometimes a request came from a representative at a global news powerhouse — a segment producer at CNN, perhaps, or a reporter at the *New York Times*. And sometimes it came from a colleague who was simply hoping for advice.

Today one such request had come from a friend at NASA. Would Heidi possibly have time to look at a Webb image and suggest how to phrase its significance for a press release?

Hammel opened the file on the computer in her home office.

She said nothing. She had nothing to say. She just sat there, staring in silence, a catch in her chest surprising her. On the screen in front of her was an image she had been hoping to see for more than three decades — nearly the entirety of her career — without knowing whether she ever would. Yet now here it was: Neptune's rings.

She blinked back a tear.

"Mom," she called upstairs to the only person within shouting distance.

Her kids were grown and out of the house. Her husband was at work.

"Mom!" she called again. "You have to see this!"

Yet even then, even as the footfalls overhead told her that another witness to the moment would soon be arriving, she couldn't wait. The image was too much to bear on her own. She swiveled her chair away from the screen, bent to the floor, and picked up the family cat. Knowing she was being ludicrous but not caring, Hammel swiveled back to the screen and hefted the cat so it was facing the screen.

"Look!" she said to the cat. "Look at the rings!"

In 1609, when Galileo began carting his *perspicillum* into his garden in Padua on a nightly basis (absence of cloud cover allowing), the universe was divided in two.

One realm was the terrestrial, from the Latin root *terra,* or earth. The other realm was the celestial, from the Latin root *caelum,* or heavens. Those two realms were all there was, the cosmos in its entirety: Earth; everything else.

That cosmic bifurcation had previously sufficed. It reflected a census of objects in the sky that hadn't changed since our species was spending its nights on the African veldt: Moon; Sun; five *planētes asters,* as the Greeks eventually called the wandering stars (Mercury, Venus, Mars, Jupiter, and Saturn); and the chorus of fixed stars that moved in unison as they circled the Earth. When Galileo first pointed his rudimentary telescope at the celestial realm, he had no reason to think that doing so would add to the cosmic population. He might very well not have found anything new in the night sky. But he had.

In early January 1610 he saw points of light on either side of Jupiter that he assumed were, like so many other points of light that had blossomed into view under the magnifying powers of his instrument, stars that were otherwise too dim to see. But over the next few evenings he noticed that these "stars" varied in number and position from night to night, and sometimes even over the course of a night. After tracking their movements for a couple of weeks, he concluded that what he was observing weren't stars but four moons orbiting Jupiter, thereby demonstrating that Earth was not at the center of all celestial rotations. Still, just because Jupiter was a center of one set of rotations didn't mean that Earth wasn't the center of rotations on the whole. The geocentric model was still intact.

That same year, though, Galileo began observing Venus, and he saw that the planet passed through phases similar to the Moon's — a procession of crescents that waxed and waned. Technically Venus could be orbiting the Sun while the Sun orbits the Earth, just as moons could be orbiting Jupiter while Jupiter orbits the Earth. But for two thousand years astronomy had accommodated similar contortions of mathematical logic, all in an effort to keep the Earth at the center of the universe. Here, though, was empirical evidence — actual observations that spanned Aristotle's previously impassable distance from *Earth* to *everything else*. The evidence was becoming more than persuasive that the model advanced by the Polish polymath Nicolaus Copernicus some seven decades earlier — a universe with the Sun at its center — wasn't just a mathematical tour de force but a template for reality.

"It was granted to me alone to discover all the new phenomena in the sky," Galileo wrote late in life, "and nothing to anybody else." Which would have been true if other astronomers hadn't continually increased the telescope's powers of magnification by adding their own innovations to the instrument — for instance, by adjusting the curvature of the lenses at either end and lengthening the tube.

In 1655, thirteen years after Galileo's death, the Dutch astronomer Christiaan Huygens used a telescope capable of fiftyfold magnification to discover another moon, this time orbiting Saturn. A year later, he used a telescope twice as powerful to solve a mystery that had enraged Galileo: "handles" that Galileo had observed disappearing and reappearing on either side of Saturn. These protrusions, Huygens wrote, were in fact a "thin, flat ring, nowhere touching" the planet itself. (The visibility of the ring depended on its orientation toward Earth.) In the decades to come, Giovanni Domenico Cassini identified four more moons of Saturn and determined that the ring around Saturn was actually *rings,* plural, separated by gaps.

A horizon had receded. Earth was no longer on its own, distinct from everything else. The term *terrestrial* still encompassed only Earth. But Earth itself was now also celestial. It was now a planet in — to invoke a term that came into usage early in the eighteenth century — the solar system. Each word in the term deserves equal emphasis: *solar,* because we all orbit the same Sun; *system,* because, according to the law of universal gravitation that Newton had introduced barely a generation earlier, in 1687, all of our motions are mutually interdependent.

By the early years of the twenty-first century, when the

decadal survey had endowed the Next Generation Space Telescope with its first formal imprimatur, the census of the solar system had expanded somewhat: nine planets rather than six; dozens of moons; more rings around planets; rings around moons, even. At that time the Next Generation Space Telescope had reached the stage of development where the eventual capabilities of a telescope are still under discussion. The telescope's targets now officially included the two that had defined the mission since the 1980s: planets around other stars in our galaxy; and other galaxies stretching as far as possible to the beginning of the universe. But now the time had come to form a Science Working Group to ask what else the telescope might do. What other science might it perform? Could it, for instance, accommodate the field that Galileo had inadvertently founded — solar-system science?

Those questions were precisely the kind that Dan Goldin had cautioned against in his "faster, better, cheaper" campaign in the 1990s. And that campaign had enjoyed some notable successes: the 1997 Mars Pathfinder mission, which delivered the rover Sojourner to the planet's surface, and the 1998 Lunar Prospector mission, which found evidence of ice water on the Moon.

But in 1999 three missions not only failed but did so for at least somewhat embarrassing reasons. The Wide-Field Infrared Explorer had a design flaw that ejected the telescope's dust shield within hours of launch. The Mars Polar Lander apparently experienced a premature shutdown of the engine during the spacecraft's approach to the planet's surface; in any case, it crashed. Most egregious was the failure of the Mars Climate Orbiter, which entered the Martian atmosphere at an incorrect trajectory because someone had forgotten to convert instructions from English units to metric units.

By 2002, Goldin was gone — a victim of the administration change in 2001 as well as the failures of 1999 — and NASA was shifting back to its old, more-bang-for-more-bucks-via-fewer-missions philosophy. When Heidi Hammel surveyed the decade ahead from the perspective of her own area of interest, she saw that no new planetary missions were under consideration. So she submitted her application to join the Science Working Group on behalf of solar-system science, not least because if the Next Generation Space Telescope couldn't accommodate the next generation of solar-system astronomers, there might not *be* a next generation of solar-system astronomers.

Broadly speaking, there are two stages in astronomy. One is "discovery" — the kind where you point the telescope at a promising target just to see what's out there. That's the type of program that new instruments and virgin territory need. That's the type of program that Galileo followed in propping up his telescope in a garden and pointing it at Jupiter, just to see what he could see (and what he saw redefined the universe). That's the kind of program that the Hubble telescope project welcomed, at least at first; it's the kind of program that the Hubble telescope followed in drilling through a Deep Field (and what it saw redefined the universe).

By now, though, Hubble's observing programs had matured out of the discovery stage and entered the next stage: hypothesis, analysis, interpretation. Now that you know what's out there, what do you want to know next? Now that you've seen the thousands of galaxies in the Hubble Deep Field, what do you want to learn from them—*specifically*? Define that objective as narrowly as possible. Explain in detail how you plan to meet your goal. Justify your time on this particular instrument.

Hammel's proposal to join the Science Working Group was successful. It was, appropriately, of the discovery variety. As she wrote: "Virtually any observations are guaranteed to break new ground." Later in the decade, nearer to launch, a committee would be assigning observing time for more specific searches. For now, though, Hammel's role as a member of the Working Group was, as she saw it, to be "a voice in the room." To go to meeting after meeting, year after year. To advocate again and again and again for an instrument that could perform the two essential tasks of solar-system science: tracking swiftly moving targets (such as comets, asteroids, moons) and looking directly at bright objects (such as Mars or the Great Red Spot of Jupiter). From her perspective, what was at stake was nothing less than the future of a field that she had helped to define.

As a graduate student at the University of Hawai'i in the mid-1980s, Hammel had deliberately chosen as her area of study the wide-open frontier of the outer planets. Uranus and Neptune were worlds so distant that astronomers were only then beginning to examine them at a level even remotely resembling "in detail." Uranus lay in space more than sixteen astronomical units (AU) from Earth—that is, about sixteen times the distance between Earth and the Sun (93 million miles), or 1.4 billion miles. Neptune, verging on twenty-nine AU, was nearly twice again as distant. Yet both planets were about to be within astronomers' reach. Voyager 2, a NASA probe that had launched in 1977, would soon be passing both Uranus and Neptune on its eventual journey beyond the solar system. You couldn't guess what Voyager 2 might discover, but you knew it would be worth the journey—a leap into the unknown that appealed to Hammel.

She was, at heart, a Deadhead. She loved going to Grateful Dead concerts. (She would meet her husband in a Deadhead chat room while they were both postdocs at MIT.) At a Grateful Dead concert you never knew where the music was going, not least because the musicians didn't know. You and the band and the crowd were just having an adventure together, trusting that you'd know where you were heading once you got there. In a word: *discovery.*

Hammel, along with other promising young outer-planet astronomers, received an invitation from the Jet Propulsion Laboratory (JPL) in Pasadena to watch the images of Uranus as they arrived from Voyager 2 in late 1985 and early 1986. Voyager 2 didn't disappoint most astronomers: It looked at the planet just to see what it could see, and it saw plenty, including eleven new moons and two new rings. It did not, however, provide much information about Hammel's particular subspecialty, planetary atmospherics.

So she had no choice but to wait for Voyager 2 to reach Neptune three years later. This time, though, she didn't want to be standing among the passive participants, watching the resident experts downloading and interpreting data. She wanted to be one of those experts herself.

Hammel returned to the University of Hawai'i and, over the following two years, used the university's 88-inch telescope on Mauna Kea to investigate Neptune's atmosphere. Almost at once she made a discovery. Astronomers who had been studying Neptune for the previous decade had seen some clouds in both the Northern and Southern hemispheres. But Hammel saw precisely one cloud. What had happened to all the others? She couldn't say.

Then she made another discovery. Partly by studying the motion of the cloud as it rotated along with the planet, she was in a position to argue that Neptune's rotation rate was faster than the one everyone else had agreed on.

By the time Voyager 2 performed its flyby of Neptune in August 1989, Hammel's reputation as an outer-planet expert had earned her the right to advance from bystander to participant. Not at JPL, alas. Her expertise as an observer on the

88-inch telescope meant that she was effectively relegated to that instrument, and thus to Hawaiʻi. There she would be collecting data that would provide context for the Voyager 2 images that JPL was downloading half an ocean away.

This time, though, an outer planet rewarded Hammel's hope for mysterious atmospherics: an Earth-sized storm that she helped name the Great Dark Spot; other storms that came and went with puzzling rapidity; fast-scudding clouds; confirmation of the existence of Neptune's hypothetical rings, which some previous observers had sworn they'd seen and other observers had sworn they couldn't find.

The innermost rings she found particularly tantalizing. They suggested the presence of dust, and the presence of dust so close to the planet's cloud cover suggested a possible influence on the atmospherics. She wanted a closer look, but for now she was willing to take what she could get. After all, she was young, and so was her career.

Thirty years passed.

By then — by the time Webb was nearing any one of its several potential launch dates in the mid-2010s, or in the late 2010s, or in the early 2020s — the solar system was several classifications richer and one planet poorer.

Astronomers had known since the late 1970s that Pluto has a moon, Charon. No surprise; planets have moons. Not until 1992, however, did astronomers discover an object occupying the same outer reaches of the solar system as Pluto and Charon but — like Pluto, Neptune, Uranus, Saturn, Jupiter, Mars, Earth, Venus, and Mercury — orbiting nothing but the Sun. Was it therefore a planet too?

The discovery of that object — Albion — validated a suspicion that astronomers had long held. Ever since Clyde Tombaugh's discovery of Pluto in 1930, astronomers had considered the ninth planet an anomaly. The existence of a solo small rock so far from the Sun — approximately forty astronomical units, or a third more distant than Neptune — defied a sort of geological logic, disrupting what otherwise seemed an orderly series. The solar system, working outward from the Sun, consisted of four rocks (Mercury, Venus, Earth, Mars), then a belt of smaller rocks (asteroids), then two gas giants (Jupiter and Saturn), then two ice giants (Uranus and Neptune). Why would a planet beyond Neptune be a rock? Even if the presence of a rock at that distance wound up making sense, why would it be the only rock out there? And then, with the discovery of Albion, astronomers knew it wasn't.

Shortly after that discovery, several astronomical collaborations, or scientific teams, began dedicating observing programs (in particular at Palomar Observatory, in the mountains northeast of San Diego) to finding more objects at a similar distance from the Sun, all orbiting the Sun and the Sun alone. Those programs were successful in finding Quaoar in 2002, Sedna in 2003, Orcus and Haumea in 2004, Eris and Makemake in 2005, and Gonggong in 2007.

Astronomers clearly needed a new designation for these discoveries, and the one they settled on was "dwarf planet" — a category to which the International Astronomical Union relegated Pluto in 2006. But also orbiting within those far reaches of the solar system, astronomers found, were thousands of less massive objects that didn't seem to rise to the level of "planet," dwarf or otherwise; astronomers designated everything in that orbital realm to be Kuiper Belt Objects, the existence of which the Dutch astronomer Gerard Kuiper had hypothesized in 1951. But what of the objects *not* traveling within the Kuiper Belt? Best to relegate *everything* on the other side of Neptune to another new category, Trans-Neptunian Objects.

Yet even as the census of solar-system objects was multiplying, the prospect that Webb might perform solar-system science was diminishing. An interim Science Working Group — the ad hoc coalition that had guided the Next Generation Space Telescope through the 2001 decadal review process — had enshrined the ability to track moving targets as a Level One requirement, meaning that the project would have to meet that goal. Requiring the telescope to be able to track moving targets was, in essence, a guarantee that it could perform solar-system astronomy. But that guarantee was one that a committee was requiring in an unofficial capacity, rendering the requirement not a requirement at all. One day a few years later, when Hammel — as a member of the "real" Science Working Group — was reviewing the latest list of Level One priorities for Webb, she noticed with a start that moving-target tracking was gone.

The omission didn't lack for logic. Neither of Webb's primary goals required the capacity to track (sometimes swiftly) moving objects. Planets around stars in the rest of our galaxy; galaxies in the early universe: Both targets were too distant to register motions necessitating rapid tracking in a telescope. Not that the absence of moving-target tracking from Level One requirements meant that in the end the mission would definitely not include moving-target tracking. But should budget cuts become a necessity, that technology would be among the first to go, taking solar-system astronomy with it.

Even the presence of on-site solar-system astronomers at the Institute was waning. Once they had been a major part of the program, during Hubble's gestation and through its maturity, but by the early 2010s their number had dwindled to one, and then he resigned too.

But then, how could you justify adding staff to a mission that was already experiencing launch delays and budget creep? Hammel and her colleagues thought they might know how: dithering.

Dithering is a digital technique that astrophotographers commonly apply to observations of galaxies: Take a number of exposures, and jostle the detector slightly between one exposure and the next. In each exposure, one or more of the pixels might be junk—for instance, a stray cosmic ray that might have emerged from the Sun but, then again, might have arrived from deep space. In reconstructing the image, though, you could replace the "bad" pixels with the "good," and the result would be a reliably complete picture.

Now Hammel and some colleagues suggested applying the same strategy not to observing distant *stationary* objects but to tracking nearby *moving* objects. Dithering, they argued, was already a task Webb would be performing. Quite so, NASA sages agreed, and thereafter solar-system astronomy remained within the capabilities of Webb. In 2012 the Institute accordingly hired a solar-system specialist, John Stansberry.

By then, Hammel herself was no longer an active astronomer. In 2010, at the age of fifty, she had accepted the position of executive vice president of AURA, the independent organization overseeing research in astronomy at numerous institutions. But as one of Webb's original interdisciplinary scientists serving on the Science Working Group, Hammel was automatically eligible for one hundred hours in the Guaranteed Time Observation program during the telescope's first year or so of operation. As the name stipulates, those hours were guaranteed. They were Hammel's to use however she pleased. She could give them away if she wanted. To anyone she chose, for that matter.

To the next generation, for instance.

In 1993 astronomers identified remnants of a comet that, having passed too close to Jupiter, had begun to disintegrate under that planet's gravitational pull. Calculations showed that the fragments of Comet Shoemaker-Levy 9 would crash into the planet in July 1994. Hammel was one of many astronomers who submitted proposals for observing time on Hubble to study the crash; NASA approved five of those proposals, rolled them into one, and appointed Hammel, to her shock, as principal investigator.

Nobody knew what to expect, if only because nobody had ever seen an extraterrestrial collision of two solar-system objects. All anyone knew was that the world, via images from the Hubble Space Telescope, would be watching.* During the months of preparation before the bombardment of comet

---

* Because Hammel had ensured that it could. She had insisted to NASA that the agency use the occasion as an educational opportunity, and she had organized media outreach accordingly. That July 16, during a press conference in the auditorium of the Space Telescope Science Institute, the discoverers of the comet—David Levy and Carolyn and Eugene M. Shoemaker—were in the middle of lowering expectations when, one floor below, Hammel and her team began seeing the Hubble images. They were vivid. They were violent. *Mind blown,* Hammel thought.

"I'm going up there," she said, referring to the auditorium and, by extension, the world.

"You can't do that," a NASA rep said.

"Get me a printout," Hammel said to an aide.

Moments later she interrupted the press conference, showing the printout to the astronomers and then turning it toward the audience. So evident was her excitement, so contagious her enthusiasm, that she wound up providing daily updates throughout the week. (In 2002 Heidi Hammel received the Sagan Medal from the American Astronomical Society in recognition of her skills communicating planetary science to the public.)

fragments commenced, Hammel approached everyone who had submitted proposals and asked what, if they were in her role, they would be hoping to discover.

We want to see super-gigantic plumes, one said.

We want to look for water, said another.

We want to perform spectroscopy, said a third.

When the weeklong rain of detritus ended on July 22, they'd all gotten their wishes. Hubble saw gigantic plumes. Hubble saw expanding rains. Hubble provided spectroscopy on the dark holes that had opened in Jupiter's atmosphere.

Now Hammel adopted the same strategy in deciding what to do with her guaranteed time on Webb. She surveyed the members of the solar-system community, especially the younger ones, asking what prospect in particular excited them.

Spectroscopy, one said.

Spectroscopy, said another.

Spectroscopy, said a third.

Not a surprise, perhaps, this focus on spectroscopy. It was always going to be part of Webb's core assets. Spectroscopy provided information that astronomers couldn't retrieve via any other means — information that for much of the history of modern astronomy had been not just unattainable but unimaginable, yet now was essential.

In the 1830s the French philosopher Auguste Comte argued that what is unknowable is not worth pursuing. To illustrate his position, he cited what he thought was an irrefutable example—the makeup of celestial bodies: "We conceive the possibility of determining their forms, their distances, their magnitudes, and their movements"—the qualities available through a telescope—"but we can never by any means investigate their chemical composition or mineralogical structure." To which he later added: "I am not afraid to assert that it will always be so."

Never. Always. Just as Aristotle and his intellectual heirs couldn't anticipate the invention of an instrument that would collapse the distance between a faraway object and your eyeball, Comte couldn't anticipate an instrument that would collapse the distance between a faraway object and your hands. Even as Comte was making his fatalistic pronouncements, however, the means for making such discoveries was already available, though nobody knew it.

In 1815 the German spectacle maker Josef von Fraunhofer reproduced Newton's prism experiment—passing sunlight through a prism to reveal a continuous spectrum of the light's otherwise invisible color composition. But Fraunhofer thought to pass the sunlight through a telescope first, before it hit a prism, and what he found was "an almost countless number of strong and weak vertical lines." The initial source was still sunlight, but the telescope was apparently revealing subtleties that were undetectable without magnification. What those subtleties might mean, though, eluded Fraunhofer.

In 1859—two years after Comte's death—a pair of

German scientists, the chemist Robert Wilhelm Bunsen and the physicist Gustav Robert Kirchhoff, finally figured out the significance of the lines: They correspond to the chemical constitution of the matter producing them. Different chemicals have different patterns, and each chemical has its own distinctive pattern. In principle, you could point a telescope at any celestial object, funnel its light through a magnifying lens, and cross the chasm that Comte had deemed forever unbridgeable. Through the science of slicing the electromagnetic spectrum into fine slivers—spectroscopy—you could determine the chemical composition of an object thousands or millions or, as twentieth-century astronomers came to appreciate, billions of light-years distant without leaving the lab.

That Webb would prioritize spectroscopy was never in doubt. Right from the start the two core goals of Webb's mission demanded as much: the atmospheres of extrasolar planets and the composition of their host stars within our own galaxy, and the chemistry, physics, and distances of the earliest galaxies. While spectroscopy was not a technique that nonscientists would necessarily know about, for astronomers the anticipation leading up to the launch of Webb was in a category all its own: The observatory would be performing spectroscopy in the near- and mid-infrared wavelengths—regimes within the electromagnetic spectrum that astronomical spectrographs had barely begun to explore, and certainly at nowhere near Webb's level of precision.

For astronomers specializing in the solar system, however,

the spectroscopic promise of Webb offered a bonus: It would be performing spectroscopy on objects that had never undergone spectroscopic analysis, if only because astronomers hadn't known those objects were out there: those asteroids, those Trans-Neptunian Objects, those Kuiper Belt Objects, those... who knew?

Talk about discovery. Hammel donated all one hundred hours of her guaranteed time. Stansberry, the solar-system astronomer who joined the Space Telescope Science Institute in 2012, donated some of his own time while coordinating a consortium of astronomers who would be studying his own area of expertise, Kuiper Belt Objects. Other astronomers coordinated their own areas of expertise so as to reduce unnecessary redundancies. The result was what Hammel called a "solar-system sampler"—a term that might suggest a catalogue, an update of a census. Which is what it was. But it was also an update with a difference. Webb would compile a catalogue not just of objects in the solar system but of their history.

That emphasis reflected a shift in how astronomers thought about the solar system. It wasn't just the census of the solar system that had changed over the previous two or three decades. It was the *sense* of the solar system.

The solar-system astronomers applying for observing time on Webb were still asking the same science-defining questions as their predecessors two centuries earlier: *What's out there? What are their motions?* But a third question was beginning to emerge. It was a question they might not have thought to ask (let alone hope to answer) without the promise of applying spectroscopy to the study of comets, asteroids, Kuiper Belt Objects, Trans-Neptunian Objects, planets, rings, moons. Not just *What's out there?* and *What are their motions?* but *What are their motions over time?*

Where were these objects before? Where are they now? How did they get here?

All of which is to say: *How did we get here?*

Scientists' ability to begin to answer that question was contingent on the same two functions that modern solar-system astronomy had always required: looking at bright objects, and tracking fast motions. Not until Webb started collecting data in the summer of 2022, though, could astronomers know just how well the telescope would be able to perform those two functions.

Imke de Pater, an astronomer at the University of California, Berkeley, and one of the foremost planetary experts on the Webb mission,* oversaw a team investigating the entire Jovian system—Jupiter and its attendant moons and

* Like Hammel, she was a principal observer on the 1994 Comet Shoemaker-Levy 9 impact on Jupiter, although instead of Hubble she was using the Keck Observatory on Mauna Kea.

rings. Her collaboration started with the planet itself, studying Jupiter at ten-hour intervals—or once every rotation—to understand the planet's wind patterns. In the process her team got a surprise of the kind that leaders of discovery programs live for: images of a high-velocity stream of clouds at Jupiter's equator.

Just as important as figuring out whether Webb could stare straight at Jupiter, however, was finding out whether Webb could stare straight *near* Jupiter, a planet that is a million times brighter than its rings. In a test image, de Pater's team saw nothing but the light from the planet—an outcome they had expected. The challenge then became to manipulate the data in such a way as to reveal what they wanted to see rather than what the planet would allow them to see.

Mark Showalter, a researcher specializing in planetary science at the SETI Institute, figured out a way to determine the pattern of scattered light from the planet—basically all the light they didn't want—and subtract it so that what remained were the rings, moons, and moonlets.

But being able to see those targets would be meaningless if Webb couldn't meet the second requirement of solar-system astronomy: tracking fast objects. Through dithering, as Hammel and others had suggested, Webb could perform that function, but the question remained: How well?

Roughly nine months after it launched, the telescope got its first big target-tracking challenge. On September 26, 2022,

NASA completed its Double Asteroid Redirection Test (DART)—a deliberate collision between a space probe and Dimorphos, a moon of the asteroid Didymos. The purpose of the mission was to determine whether a spacecraft could serve as a defense—a means of deflection—should we ever discover that a solar-system object is on a collision course with Earth. A camera aboard the DART probe would be recording the crash, as would Hubble, but some solar-system astronomers sensed an opportunity to test the limits of Webb's tracking capabilities. The experiment paid off: Webb managed to keep up with Dimorphos and Didymos much of the way, losing sight of the objects only shortly before impact. At that point Webb was tracking three times as fast as its technical allowance.

"We're never going to go that fast again," a NASA official told Hammel.

"It's fine," she said. "I get it." And she did. But now she knew that, if need be, you could push Webb at least a little—knowledge that might be useful for astronomers needing to track the objects that hold the secrets of the solar system's origins.

Water, water, everywhere.

At first Geronimo Villanueva, a planetary scientist at the Goddard Space Flight Center, didn't believe the data. Certainly his team had expected to find water. The main reason they had targeted Enceladus, a moon of Saturn, was that they knew they'd find water there. In 2005 the Cassini spacecraft had flown through a plume of water that astronomers hadn't known existed. It was shooting out of cracks in the surface of Enceladus, near its south pole, suggesting that the water originated in an ocean under the moon's surface of (as Cassini soon determined) ice. Astronomers then directed Cassini to fly through the plume six more times over the course of the spacecraft's thirteen-year mission. The plume, they concluded, was maybe a couple of hundred kilometers (or miles) long; Enceladus itself had a diameter of about five hundred kilometers (three hundred miles). So the plume stretched a distance that was a significant fraction of Enceladus's diameter. It was a big plume.

What Villanueva's team found, though, put that estimate to shame. Over the course of a four-and-a-half-minute exposure in November 2022, Webb's Near-Infrared Spectrograph measured the concentration of H_2O to determine how far the plume extended into space. Villanueva's team concluded that its reach from the surface of Enceladus into the universe was at least ten thousand kilometers—twenty times the diameter of Enceladus. In Earth terms, it was as if Old Faithful were shooting two-thirds of the way to our Moon.

The observational program wasn't a success, at least on its own terms. The astronomers didn't detect the same conducive-to-life chemicals that Cassini had found—methane, carbon dioxide, ammonia. Nor did they detect a few other chemicals that they had selected for potential spectroscopic identification—specifically, carbon monoxide, ethane, and methanol. In that sense, then, the result was somewhat disappointing. Still, the lack of detection wasn't surprising: Cassini had enjoyed the advantage of scavenging actual particles while passing through the plume; Webb was merely squinting from hundreds of millions of miles away.

But it was squinting at a field of view far broader than Cassini could encompass. And that broader field allowed Webb not only to measure the plume of water to its full extent but to survey a vast area of space around the plume. Most important, though, was that Webb was squeezing out those measurements via the on-board instrument that could provide a spectrum for every pixel in the image. And the spectra from those pixels, when Villanueva and the rest of the team saw them, were what gave the members of the collaboration pause.

Every pixel contained H_2O.

Every pixel.

Water wasn't just in the plume, and it wasn't just misting the immediate vicinity of the plume. Water was indeed everywhere, meaning that Enceladus wasn't simply spewing. It was spawning. It was defining the atmosphere of the planet itself—and when the team compared the outgassing rate (the volume of ejecta over a certain period of time) with Cassini data from fifteen years earlier, they found that the rate hadn't varied. Enceladus was emptying its internal ocean

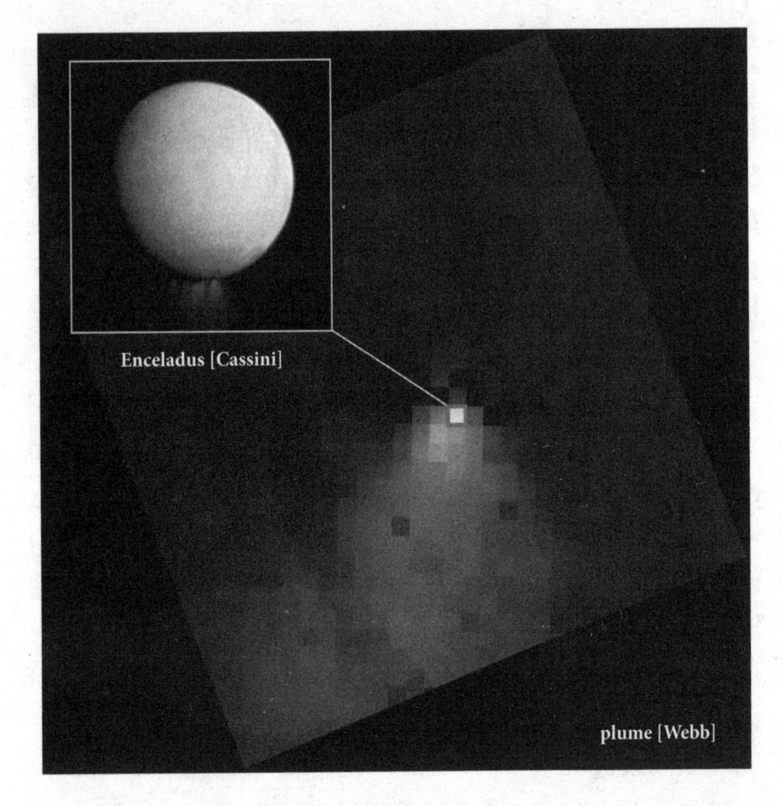

Water, water, everywhere: [Inset] In 2005 the Cassini probe discovered a geyser erupting from cracks near the south pole of Enceladus, an ice-covered moon of Saturn. [Main image] In 2022 Webb captured a more expansive view of the same phenomenon. Judging from the Cassini data, astronomers thought the geyser might reach a height of hundreds of kilometers. Webb, however, found that it not only reached at least ten thousand kilometers but was so extensive that it was influencing the atmosphere of Saturn and its rings and moons.

---

into Saturn's environs at a rate and in a volume that was affecting everything in its neighborhood. Enceladus was, as the collaboration's paper in *Nature Astronomy* reported, "the

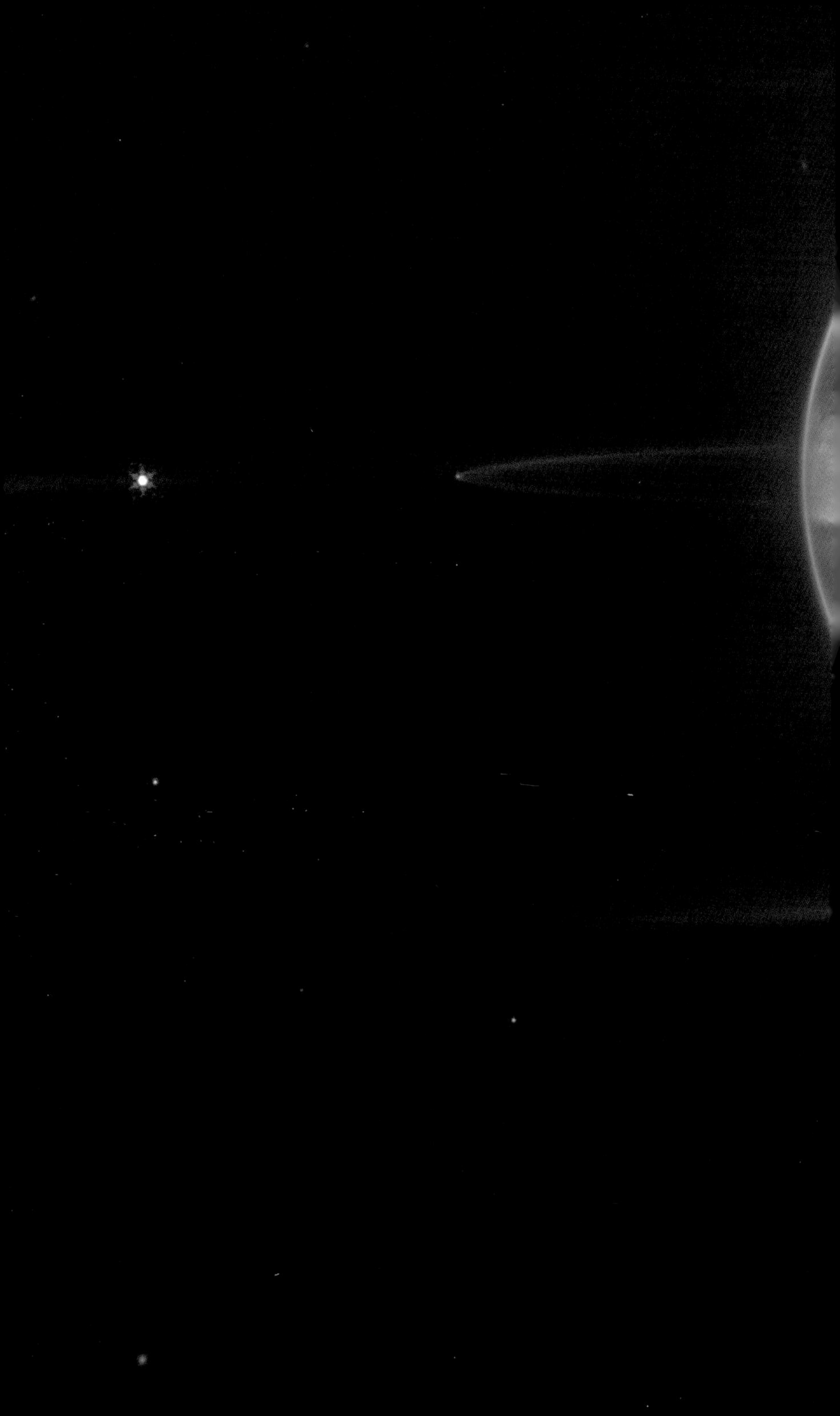

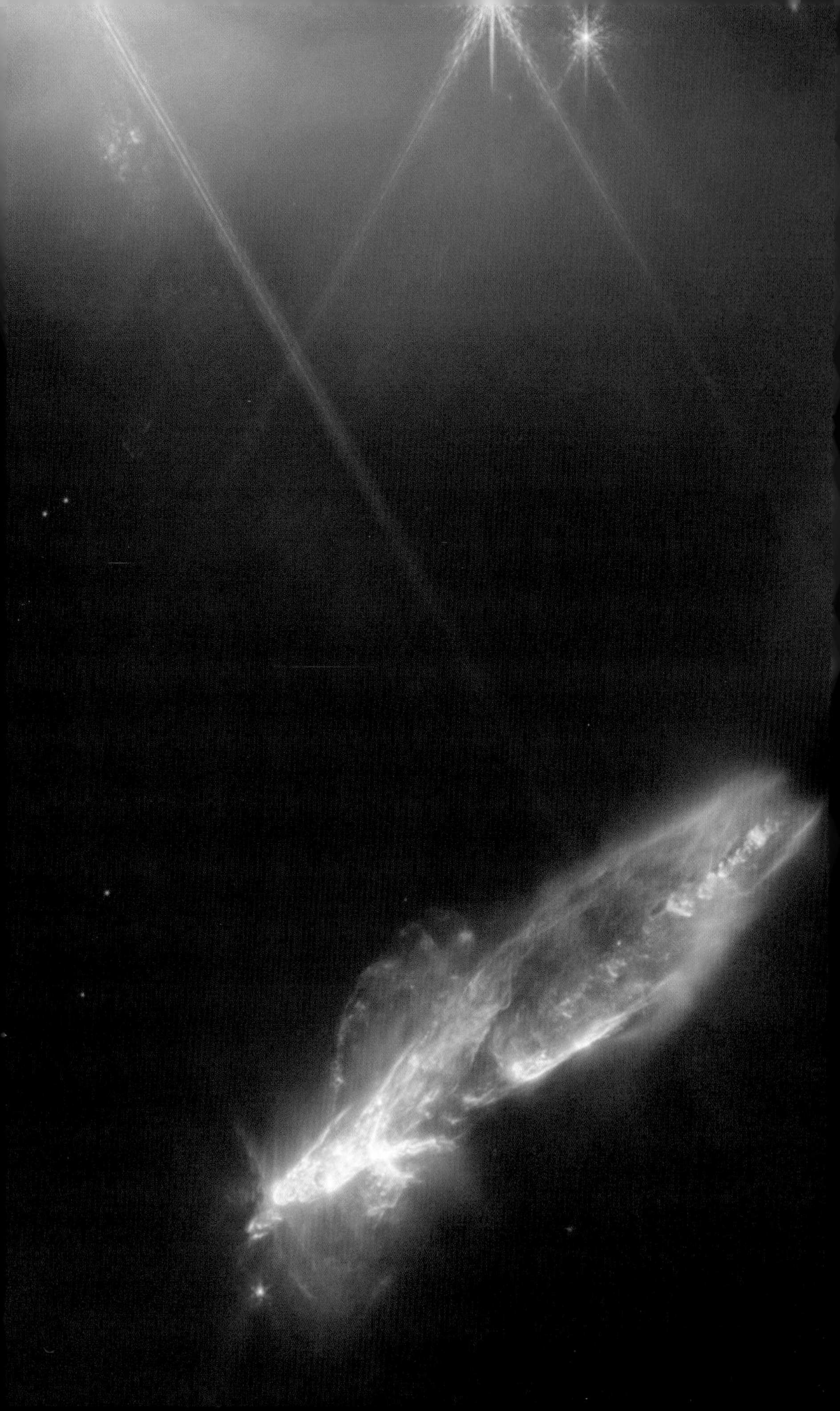

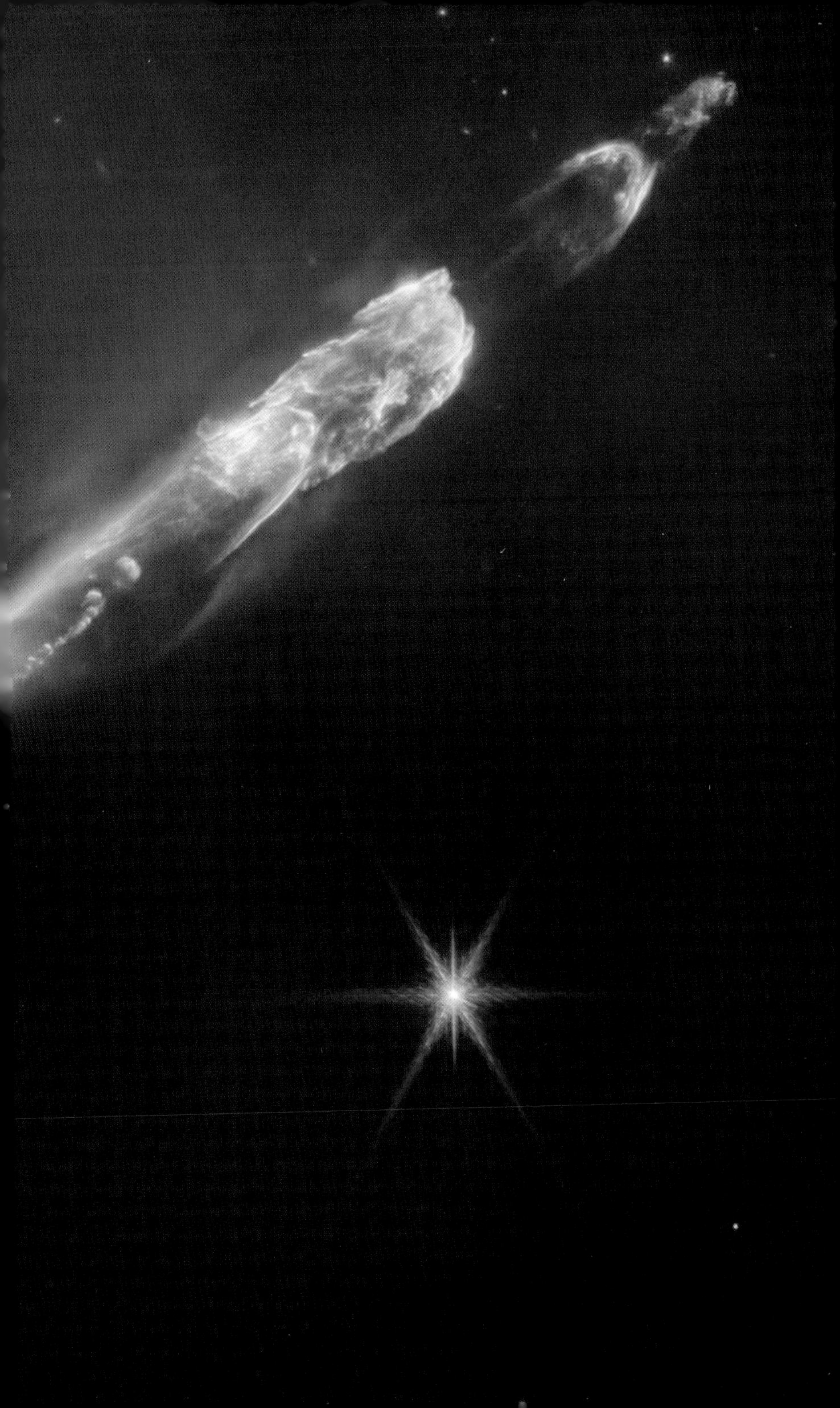

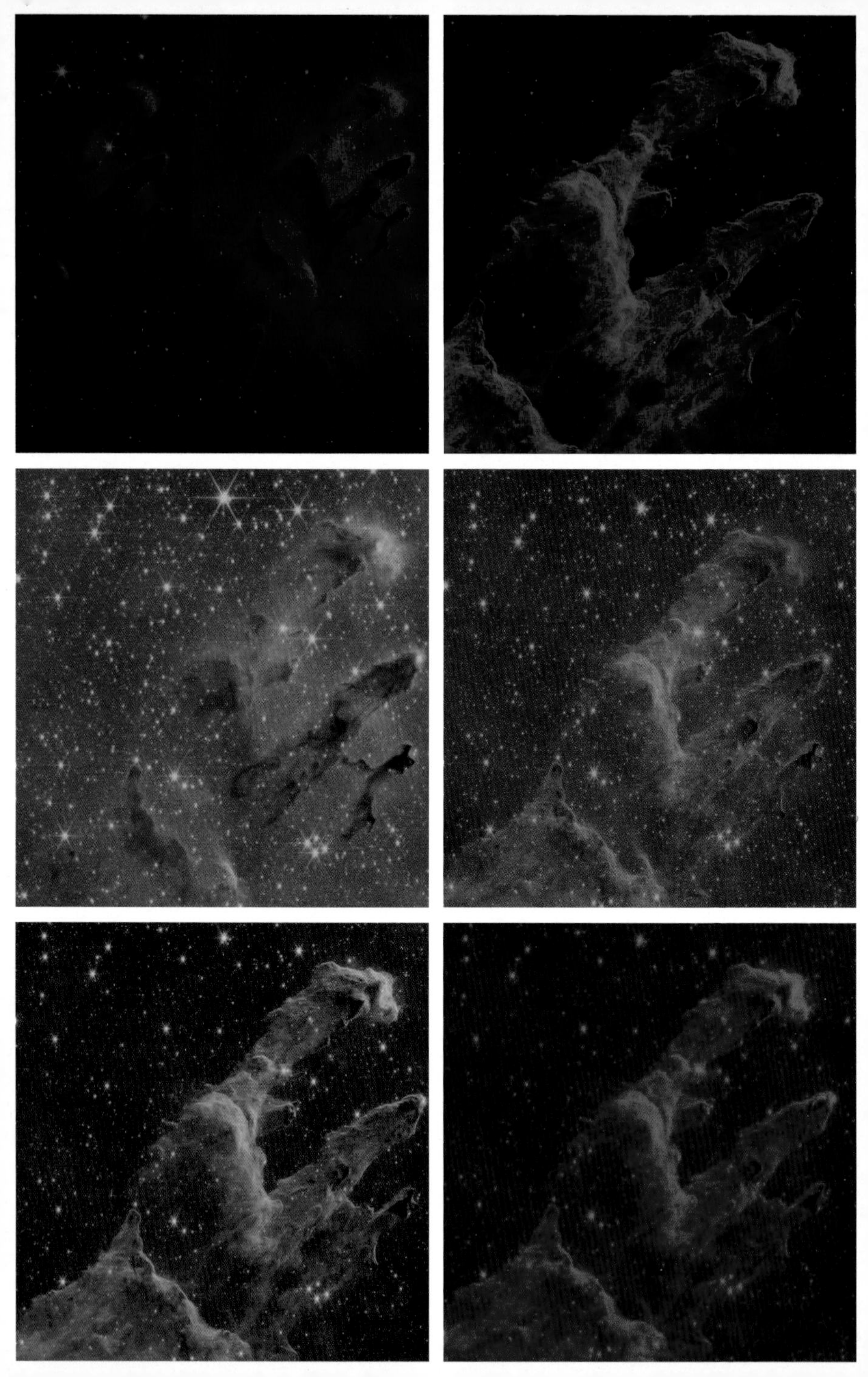

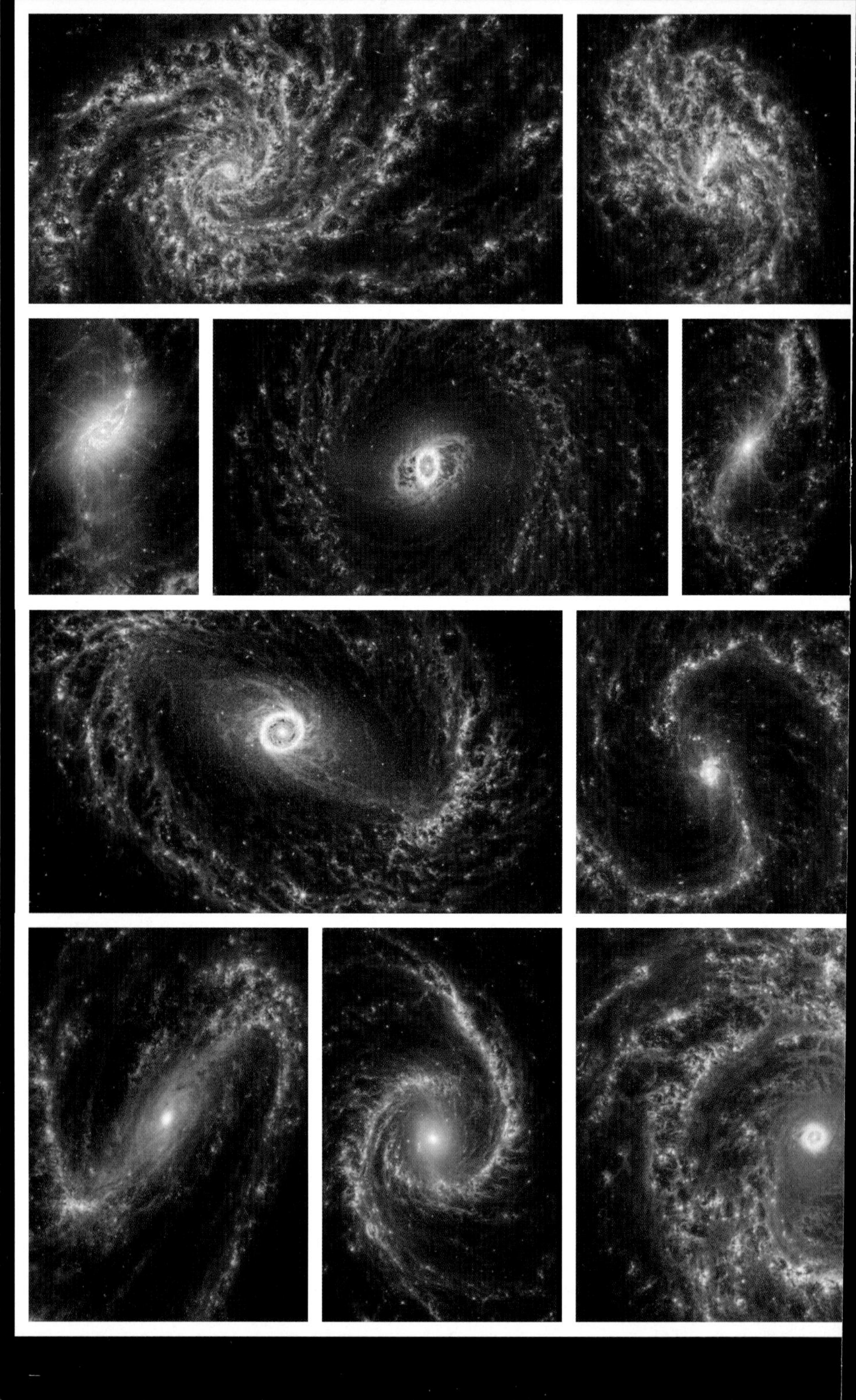

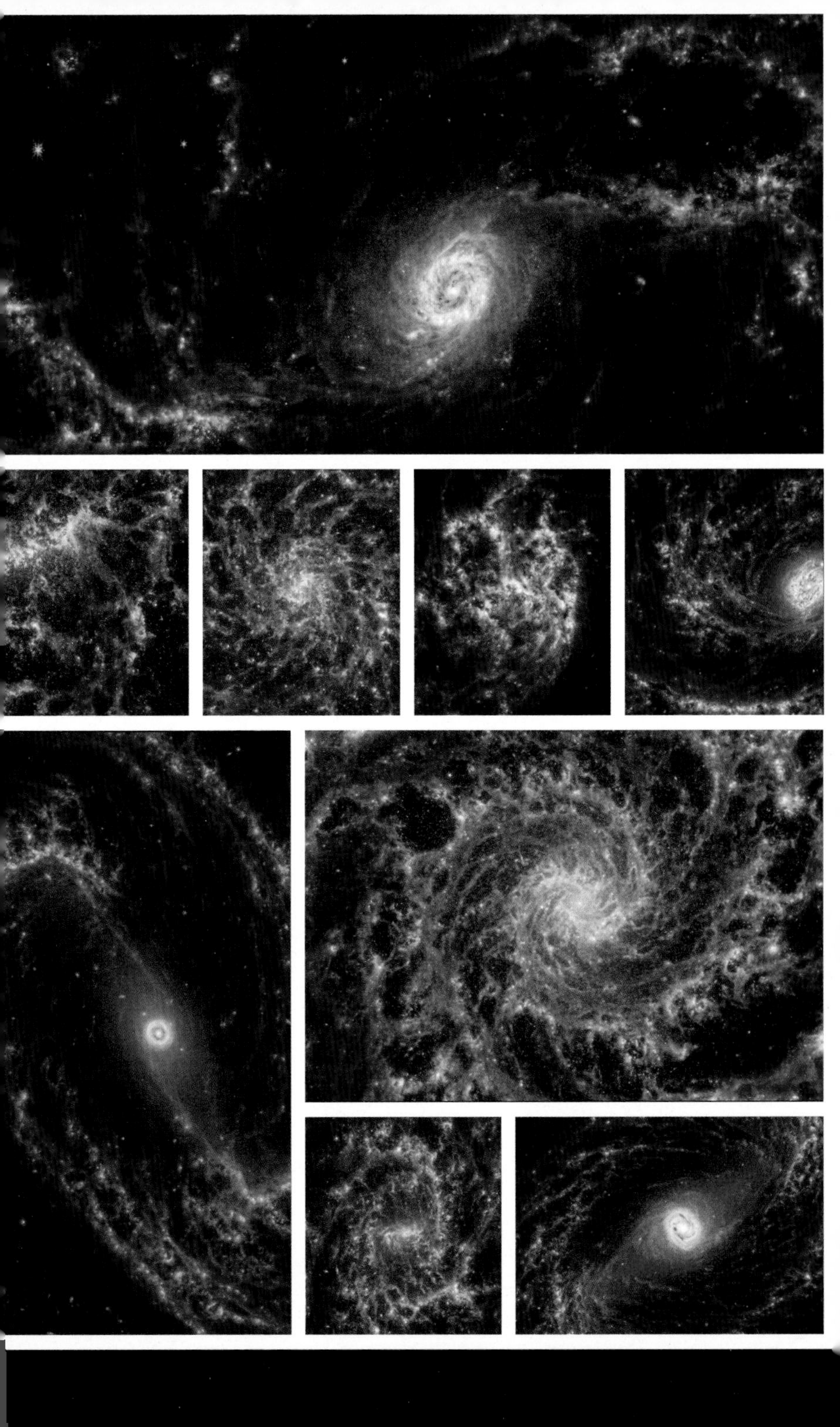

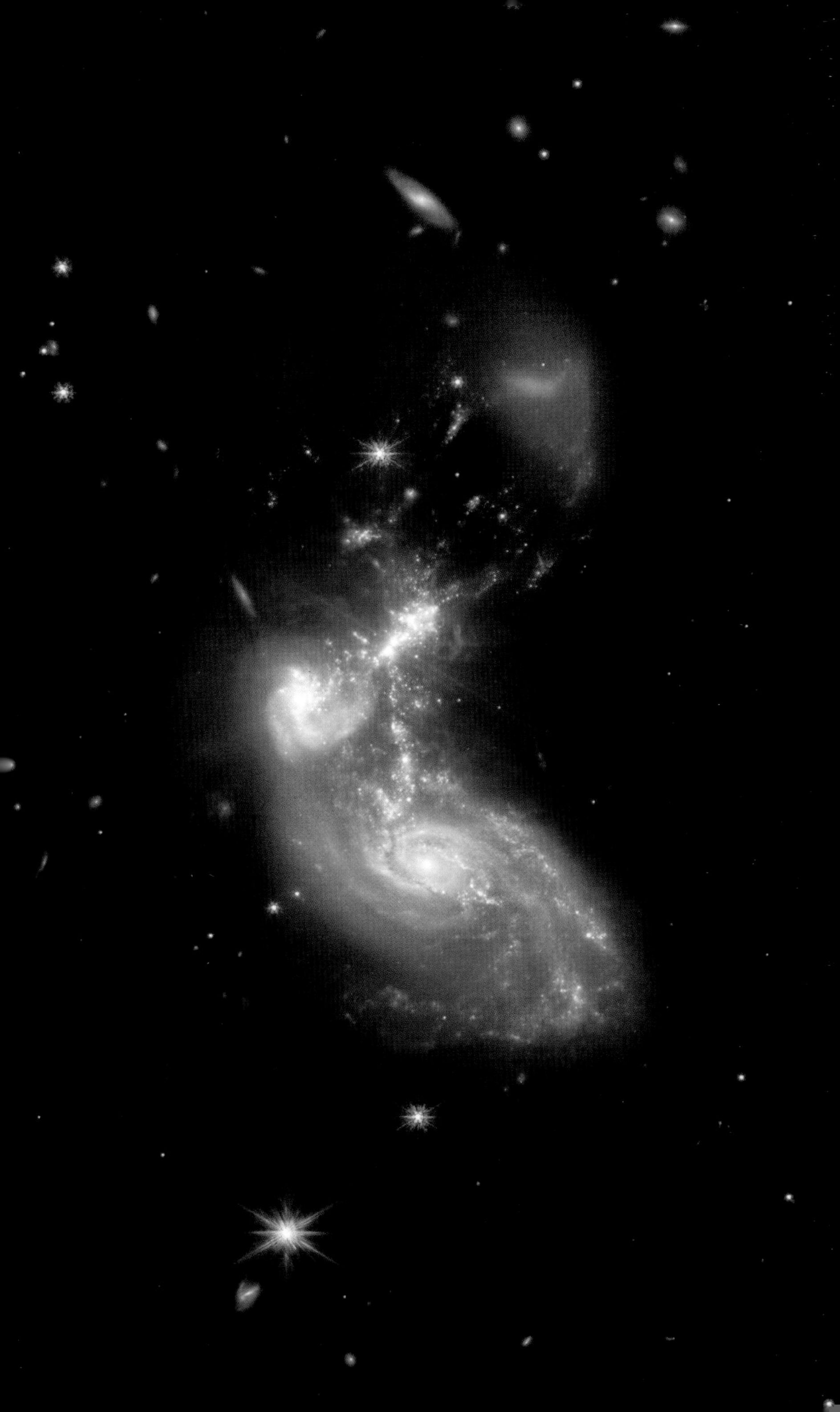

prime source of water across the Saturnian system"—meaning the planet itself along with its rings and moons.

Experts in planetary atmospherics would have to adjust their thinking about the conditions of Saturn, and they'd have to wonder whether the lessons of Enceladus, whatever they turned out to be, might apply elsewhere. The science was just beginning. As Villanueva told *Nature* magazine in an article about his team's discoveries, "This is definitely a new era in the exploration of the Solar System."

Yet as surprising as these findings were in terms of one moon's contribution to its host planet's ecosystem, they were consistent with the study of the solar system on the whole. Water, water, is indeed everywhere. It's on other planets. It's on the surface of comets. It's under the surface of moons. And it's on Earth, where life couldn't have arisen without it.

The origin of water on Earth—and with it the origin of life—was one of the major mysteries that Webb's solar-system science was to address. The longstanding working assumption had been that early in the solar system's development, comets bombarding our planet seeded it with their own water. And Webb hadn't challenged that assumption—but it *had* caused astronomers to rethink a key aspect of the evolution of the solar system.

A comet has a tail because the ice on its surface turns into gas as it nears the Sun, a process that astronomers call sublimation. The liquid in that ice needn't be water. It could be, for instance, carbon monoxide or carbon dioxide. Whatever the ice contains, astronomers know at what distance from the Sun—the "ice line"—that particular component will start sublimating.

And now, thanks to the precision and spectroscopic subtlety of Webb (as well as its ability to track fast-moving objects), astronomers began to trace that journey through the solar system. Did the liquid in the ice exist only on the surface, or was it also emerging from the comet's interior?

But even as they began to gather the information that they hoped would help them answer these kinds of basic questions, scientists were also finding anomalies that were leading to new questions. They found comets that were sublimating at ice lines farther from the Sun than astronomers thought possible. Why? They didn't know. What happens in the outer solar system, whence comets emerge on their periodic slingshot trajectories around the Sun? What *are* comets, anyway?

And what are comets, or at least "comets," doing in the asteroid belt? Asteroids, astronomers had every reason to believe, were remnants of a would-be planet that, under gravitational duress from its two neighbors—Mars on one side and, far more significantly, Jupiter on the other—never quite coalesced. When the Italian priest and astronomer Giuseppe Piazzi discovered an object between the orbits of Mars and Jupiter in 1801, he assumed it was a planet. Over the following century, several other "planet" discoveries followed. By the 1920s, the number of asteroids numbered a thousand; by the 1980s, ten thousand; by the time Webb starting studying them, a hundred thousand.

Of which about a dozen (to that point) didn't fit the standard description. Asteroids are rocks that have occupied a position in the solar system that, unlike the position of comets, hasn't varied for billions of years. At their distance from the Sun, any surface coating would have evaporated eons ago — or so you might think. Yet these dozen or so members of the asteroid belt were exhibiting a fuzzy coma — the halo of gas and dust common to comets. Some astronomers called them "active asteroids." Others preferred "main belt comets." Either way, they were sublimating. But sublimating what?

Water, Webb found — the discovery that Hammel, for one, ranked as perhaps the most significant in Webb's first two years collecting solar-system data.

But how can these objects contain water if they've been occupying this barren zone for billions of years? Asteroids, astronomers had thought, make up one of the few classes of objects in the solar system that *don't* contain water. But if even a few objects in the asteroid belt contain water *now,* did most of them, or even all of them, contain water way back when the solar system was forming? Could asteroids have accompanied comets in delivering water to Earth? If so, does the presence of water at that distance from the Sun indicate where the habitable zones of planets in general might be — the distance from their host stars at which life might exist? And in that case, what hints can these main-belt comets or active asteroids offer to astronomers who are investigating what lies beyond Webb's next horizon—the stars and planets outside our solar system, populating the rest of our Milky Way galaxy?

One image that was not a product of the "solar-system sampler" project was the photograph of Neptune that popped into Heidi Hammel's inbox on September 20, 2022. That photo was instead a figment of a NASA public-relations campaign, a composite of pretty pictures in a point-and-shoot

The image that made an astronomer blink back a tear. When Heidi Hammel first opened this file on her computer, she realized she had fulfilled a career-long wish to see Neptune's rings in detail.

operation. From a scientific perspective, the image held questionable value. It came with no spectroscopy, for instance. Scientists who knew better, like John Stansberry, shrugged.

Hammel knew better—or would have known better if she'd had more time to reflect as a scientist. Instead, she encountered the image on the visceral level NASA intended—as a civilian.

How beautiful, she thought. *How awesome.*

Then her brain made the transition, and she was a scientist again. But she still couldn't gather her composure, she still couldn't regain the detachment of a neutral scientific observer. Not entirely, anyway.

Even when, eventually, she recognized that the image held little immediate scientific value, she still appreciated its potential significance: certainly what Webb might do for the study of the influence of a planet's innermost rings on its atmospherics, but also what this instrument might do for solar-system astronomy.

The photo was, in short, her kind of science. Somebody had had an adventure. They'd pointed a telescope, trusting they would know what they were searching for once they saw it. And then they'd found something that Heidi Hammel, for thirty-three years, had been hoping she might live long enough to see again.

What a long strange trip it'd been.

CHAPTER FOUR

SECOND HORIZON: CLOSE TO HOMES?

"Hold on to your seats."

Nikku Madhusudhan, an astronomer at the University of Cambridge, had just talked his audience through some of his team's recent Webb observations of K2-18 b, an exoplanet—a planet outside our own solar system—approximately 124 light-years from Earth in the constellation Leo. He could exit now, he said. He'd already announced a couple of possibly significant findings. At this point in a conference talk, a speaker might simply cede the stage and take a victory lap around the cafeteria during the next coffee break.

But, he said, he had one more possible discovery about K2-18 b to discuss. It was something his team had found in the planet's atmosphere: possible evidence of dimethyl sulfide, a molecule that would be a biomarker — a sign of life.

The only scientifically appropriate response to such a claim is extreme caution, and Madhusudhan was quick to issue caveats. His team's confidence level, he said, was lower than it was for the other findings he'd just reported. It was certainly far too low for his team to claim a discovery.

"We don't know if this is real yet," he said. "But," he went on, "on Earth dimethyl sulfide is produced just by life. You cannot produce it any other way."

Before claiming a discovery, as he assured his audience, his team would have to perform more observations. Still, the potential significance — you couldn't overstate it: "There is something there. We don't know, so we are calling it tentative. But folks, if in a few years we realize this is actually dimethyl sulfide, then history is being made as we speak right now" — September 11, 2023, the opening day of the "First Year of JWST Science" conference at the Space Telescope Science Institute.

"I kid you not," he added.

For nearly two centuries after the invention of the telescope, the universe was divided in three: Earth; then the neighboring planets and moons in what we had become comfortable calling the solar system; then the starry realm. But that third component—the celestial vault; the firmament of the heavens—remained an astronomical afterthought, and for a good reason: Beyond the outermost edge of our solar system, a horizon defined by the orbit of Saturn and its accompanying moons and rings, lay only pinpoints of light that, even under the greatest available powers of magnification, stubbornly remained pinpoints of light.

Not that astronomers entirely ignored the stars. In 1718 Edmond Halley reported that he'd detected "proper motion" in three stars—motions indicating that while the stars on the whole appeared to move as a single unit, at least a few were moving among themselves. Not long after, John Flamsteed, England's first Astronomer Royal, compiled a catalogue of three thousand stars, the 1725 *Historia Coelestis Britannica*. By the end of that decade the English astronomer James Bradley had calculated that the distance from Earth to the nearest star must be at least 400,000 astronomical units, or 36 trillion miles.

The nearest star: The conceptual breakthrough was almost there. The existence of a nearest star implies the existence of farther stars. But just as Galileo wasn't going to see new phenomena in the night sky as long as he was using his version of the telescope rather than the potentially more powerful one that Kepler had invented, so astronomers weren't going to explore the realm beyond the solar system as long as they were using the Keplerian version of the telescope.

An alternative *was* out there. It just didn't seem to offer an advantage over Kepler's version. The Keplerian model of telescope relied on *refraction*—the redirection of light as it passed through the lens at the far end of the instrument before coming to a focus near a second lens at the base. An alternative, though, had arisen in the 1660s and 1670s through the ingenuity of several inventors, including Isaac Newton. This version relied on *reflection.* Light still entered at the sky-facing end, but it didn't pass through a lens. Instead, it simply raced down the shaft without interruption, at least until it reached a paraboloid mirror at the base, at which point it bounced back to a tiny mirror midway down the shaft, at which point it bounced again, this time toward a lens at the side, where an eyeball awaited. But all the eyeball saw was an image with a magnification and level of detail comparable to what a Keplerian telescope would have delivered—so why bother?

When the German-English amateur astronomer (and professional church organist) William Herschel began his own observations in the 1770s, he used the instrument that he'd read all the real astronomers were using—the refracting telescope. But as he turned to longer and longer telescopes—twelve feet, fifteen feet, thirty feet—he found them so increasingly cumbersome that he resorted to renting a reflector. It provided results similar to those of his refractors, but, at a length of only two feet, it offered far easier maneuverability.

Soon, though, Herschel realized that the reflecting telescope also possessed a quality that professional astronomers hadn't appreciated: It could see farther than a refractor.

Herschel found that the reflecting telescope follows a simple logical progression. First, a mirror captures more light than a lens of the same diameter. Second, the greater the diameter of a mirror, the greater the amount of light it captures. Third, the greater the amount of light, the deeper the reach across the universe.

"The great end in view," Herschel wrote to a friend in 1785, "is to increase what I have called *the power of extending into space*." Seeing farther might not help much in studying the solar system, but among the stars? Herschel taught himself how to grind mirrors, and then he taught himself how to grind mirrors of greater and greater diameters, with each improvement in diameter dramatically deepening his view: stars fainter and fainter, stars farther and farther. His serendipitous discovery, on March 13, 1781, of the planet Uranus had nothing to do with his stellar investigations, other than that it occurred while he was conducting them. But it did double the diameter of the solar system, creating a sensation among professional astronomers and crowned heads and forcing a reconsideration of the utility of the reflecting telescope. Herschel soon found himself in the business of manufacturing mirrors for observatories and royalty, and eventually those orders reached into the hundreds. For his own purposes he produced mirrors of greater and greater diameters, including one that reached forty-eight inches. Flamsteed's 1720s catalogue of stars, a copy of which Herschel owned, consisted of three thousand entries; in one forty-one-minute observing run, Herschel estimated, he himself saw 258,000 stars.

"Hitherto the sidereal [starry] heavens have, not inadequately for the purpose designed, been represented by the concave surface of a sphere," he wrote in the mid-1780s. "The construction of the heavens," however, "can only be delineated with precision, when we have the situation of each heavenly body assigned in three dimensions, which in the case of the visible universe may be called length, breadth, and depth."

Coelorum perrupit claustra reads the epitaph on his tombstone in a parish church in Slough, England, where Herschel had lived most of his life: "He broke through the barriers of the heavens." What waited on the far side of those barriers — that horizon — was the second of the four realms that Webb began exploring in 2022, on the two hundredth anniversary of Herschel's death: stars and their planets.

In the days after Nikku Madhusudhan advised the Space Telescope Science Institute audience that they'd be wise to hold on to their seats, he began having second thoughts.

Not about the possible presence of dimethyl sulfide in the atmosphere of K2-18 b. That "detection" he had deliberately cautioned was "tentative." In private conversation, he was now calling it "very, very tentative," just to be clear about his intentions.

And not because he was wrong to assert the potential momentousness to his audience. If the detection of this molecule turned out to be valid, then yes, the people in the auditorium had indeed been witnessing history.

Instead, whatever regrets Madhusudhan harbored were in regard to how he had handled the announcement—the "I kid you not" kind of showmanship, or at least emphasis. Too late now to do anything about it, though. *Here we are,* he told himself when he considered the PR repercussions. And where they were—where the response to his team's results was—was the popular-science equivalent of a frenzy. The day of the announcement, blogs and mainstream media outlets had pounced upon the "news."

Some did so with caution: HAS LIFE BEEN DISCOVERED ON AN EXOPLANET?*

Others did so with abandon: WE JUST FOUND A MOLECULE ON ANOTHER WORLD† . . . AND ONLY LIVING ORGANISMS CAN PRODUCE IT.‡

And a few were the kind that could prompt a scientist to second-guess his manner of public presentation: BIG DISCOVERY: JAMES WEBB FOUND SIGNS OF LIFE ON K2-18 B!

---

* No. But thanks for asking.
† We did not.
‡ As far as we know, but what we know comes only from knowledge of conditions on Earth.

The "life narrative," as Madhusudhan had now come to think of it, might reside more powerfully in the human imagination than he had anticipated. And he *had* anticipated it, at least to a degree. He'd timed the online posting of his team's paper as well as the availability of a press release to coincide with the opening day of the "First Year of JWST Science" conference — the same day he gave his talk.

"The search for habitable environments and biomarkers in exoplanetary atmospheres," read the opening sentence of the paper's abstract, "is the holy grail of exoplanet science." However unfortunate the syntax (the *search* isn't the holy grail; the *discovery* is), readers were sure to get the point:

Life.

The existence of it. The potential for the existence of it. The conditions conducive to the potential existence of it. The emergence of the conditions conducive to the potential existence of it.

When astronomers speak of extraterrestrial life, they don't mean Uncle Martin or E.T., Klaatu or Klingons. They mean, maybe, life on a unicellular level. Better yet, a multicellular level. None of which obviates the possibility of more advanced forms. But the chance of stumbling across sentient beings is far, far lower than of finding the potential for the eventual existence of microbes.

Within our solar system, Webb astronomers' search for the origins of life had primarily been converging on two questions: *Where is the water?* and *How did it get here?* In the rest of the galaxy beyond our solar system, Webb astronomers' search for the origins of life was primarily converging on two similar questions: *Do the preconditions for life (presumably including water) exist?* and *If so, how did they get there?* To try to find the answers, Webb investigators were focusing on three previously obscure, if not observationally inaccessible, stages in the production of exoplanets—processes shrouded in the gas and dust that Webb's infrared sight could penetrate.

First, protostars—objects that haven't yet achieved starhood but are on their way.

In November 2022 Webb captured an image of L1527, a dark cloud about 460 light-years from Earth in the Taurus constellation. Astronomers consider L1527 a class 0 protostar, the earliest stage of star formation. As was true for many of Webb's images, the dust in L1527 was impenetrable in optical wavelengths but blazingly transparent in the infrared. What Webb revealed was an hourglass shape: two gigantic bulbs of gas and dust ballooning on either side of a narrow "neck," at the center of which resided a protostar.

The protostar itself wasn't visible, but its influence on its surroundings was. Those twin bulbs suggesting an hourglass shape were material that the protostar was ejecting, clearing space for its own eventual birth. The birth process itself was visible in the neck of the hourglass. Peer closely enough, which Webb could do, and you'd see gas and dust accreting into a disk in the immediate vicinity of the protostar. That disk, astronomers calculated, was similar in size to our own solar system. In that case, maybe the protoplanetary disk in L1527 was an analog of the process attending the birth of our Sun and our planets and our Earth.

Astronomers using other instruments had estimated that L1527 was only a hundred thousand years old. Our solar system is about 4.5 billion years old. On the scale of a human eighty-year life span, Webb astronomers were studying L1527 for clues as to how our own solar system was behaving when it was the human equivalent of thirteen hours old, back when its central disk of gas and dust was becoming more and more dense and hotter and hotter, but before gravitational compression sparked hydrogen fusion in its core and our star was born.

Water, water, still everywhere.

The second stage in the search for signs of life was the formation of protoplanets and protoplanetary systems. While many astronomers study our solar system for clues as to where the water is and how it got there, so exoplanet experts want to know what role water plays in the rest of the galaxy.

As was the case in solar-system astronomy, theorists proposed that water arrives from the farthest reaches of a protoplanetary disk — the same region where comets originate in our solar system. Specifically, according to the theory, icy pebbles in the cold, outer regions of protoplanetary disks would drift inward toward the star, carrying the solids and liquids that, in sufficient numbers, would eventually coalesce to form planets. If this scenario is correct, astronomers could expect to find large amounts of water vapor in the region where the hypothetical infalling pellets would pass the "snow line"—the point where the warmth of the host star turns ice to water vapor.

In November 2023 astronomers announced that they had found supporting evidence for that theory. Using the Medium-Resolution Spectrometer, a component of Webb's Mid-Infrared Instrument that is particularly sensitive to the detection of water vapor, they had studied four protoplanetary disks around Sun-like stars. Two of the disks were of the "compact" variety, similar in size to our solar system. The collaborators expected to find an excess of cold water vapor in those disks, since the protoplanets in those disks would be near to where the snow line is in our solar system. The two other disks were of the "extended" variety; the protoplanets within those disks lay far beyond the snow line for their Sun-like stars. The astronomers found what they'd hoped to find: a surfeit of water vapor within the compact protoplanetary disks, and a relative absence of water vapor within the extended disks.

In astronomy, though, "ice" refers not just to the solid form of a chemical compound consisting of two atoms of hydrogen and one of oxygen but to solid forms of elements—hydrogen, oxygen, carbon, nitrogen, and sulfur—and the compounds they form. Within the first six months of Webb science, an international team of astronomers studied the Chamaeleon I molecular cloud, about 630 light-years from Earth, conducting the most extensive survey to date of ices in a protoplanetary disk. Among the molecules they found were solid forms of carbon dioxide, ammonia, methane, and methanol—strong evidence supporting the idea that the existence of complex molecules pre-dates even the protoplanetary stage of stellar evolution.

But just because these ice pellets bear the prospects for life doesn't mean that they will lead to life forms. Nor does the existence of ice pellets guarantee the emergence of larger objects. If they do emerge, though, they would constitute the third stage in the search for signs of life in our galaxy, after protostars and protoplanets: the exoplanets themselves.

On a practical level, you can't really begin a search for life on exoplanets without knowing that exoplanets exist.

That they do was almost a foregone conclusion—hence their inclusion right from the start of the project that would become Webb, back in the 1980s, under the assumption that the discovery of exoplanets would come in time for the telescope to study them.

The first discovery of a planet orbiting a star other than our own came in October 1995*—coincidentally, about

* The first confirmation of the discovery of exoplanets came in 1992, but they were orbiting a pulsar—a neutron star (a hyperdense remnant of a

midway through the deliberations of Dressler's HST & Beyond Committee. The Swiss astronomers Michel Mayor and Didier Queloz reported that they had identified a Jupiter-sized planet in a four-day orbit around the star 51 Pegasi. The brevity of that orbital period suggests that the planet is moving extremely fast, which in turn suggests, according to the laws of gravity, that it's extremely close to the star.

When we think about the gravitational interaction between a star and a planet — for instance, the Sun and the Earth — we tend to emphasize only one side of the interaction: the star's effect on the planet, lassoing it throughout its orbit. But as in all things gravitational, the interaction goes both ways: The planet has an effect on the star, too.* Mayor and Queloz made their discovery with a spectrograph on a French telescope that measured the planet's influence on the star — the shortening (blueshifting) or stretching (redshifting) of the wavelengths in the light from the star 51 Pegasi as it wobbled away from or toward their telescope in its mutual interaction with the planet.

The radial-velocity method of exoplanet detection predominated for the next fifteen years. This method, however, favored the kinds of planets that could make a star's wobble detectable from Earth: Jupiter-sized giants, for instance,

---

supergiant star) rotating at a rate of just over once a second. That finding was important in a proof-of-concept way, but it was not necessarily indicative of what we might discover around a star similar to our own.

* The Earth, for instance, tugs the Sun about 9 centimeters, or about 3.5 inches, toward itself throughout its orbit.

orbiting in close proximity to their host stars. And a planet that close to its star would be too hot to harbor life. But the instrumentation that Webb's engineers were designing in the first decade of the twenty-first century would be employing an alternate method then just coming into use.

Kepler, a space telescope that launched in 2009, inaugurated a new era in exoplanet discovery. Rather than observe a star's motions, Kepler recorded the changes in a star's light as an exoplanet passed in front of—or transited—the star's surface (from our perspective). Like the radial-velocity method, the transit method was indirect. It didn't involve the actual observation of an exoplanet, only that object's effect on the light from the star "behind" it. But the transit method did allow for the detection of planets similar in size to Earth and farther from their host stars than previous exoplanets. By the time the mission ended, in 2018, Kepler had identified 2,600 exoplanets in total.

That same year, NASA launched a follow-up spacecraft, the Transiting Exoplanet Survey Satellite (TESS), which also used the transit method. In both cases—Kepler and TESS—the purpose was to count the exoplanets and characterize their masses, sizes, densities, distances from the stars, and orbital periods. By the time of Webb's launch the number of known exoplanets exceeded five thousand, of which more than four thousand belonged to a multiplanet system, while the number of known Earth-like exoplanets had reached two hundred, some three dozen of which possessed what astronomers call a "potential for habitability."

The primary criterion in identifying whether a planet is in the habitable zone is its Goldilocks-style distance from the host star. Is it too close to the star, and therefore too hot, to harbor the liquid water that we assume is a precondition for life? Is it too far from the star, and therefore too cold? Or is it just right—or right enough, anyway?

Of course, being in the habitable zone doesn't in itself promise the presence of water. Of the four planets in the Sun's habitable zone, three seem to possess evidence for the presence of water at some point in their history: Mercury, Earth, Mars. Venus, however, does not.

Then again, the presence of water doesn't in itself promise the presence of life. Mars seems to have the remnants of a seabed, its water long gone; it might also have underground ice near its equator. Mercury harbors traces of ice in its polar regions, as does our Moon. But as far as we know, only one occupant of the Sun's habitable zone has given rise to life: Earth.

Even so, the idea that water is necessary for the rise of life is only an assumption (a sound one, but still). Water *is* necessary for life as we know it, but life as we know it is all we know about life. Scientists call this dynamic the $n = 1$ problem; the only data point is one specimen.

But in the search for extraterrestrial life, identifying an Earth-like exoplanet in a habitable zone is at least a start—and making "a start" is what the discovery stage in a scientific search is for. Like Galileo in his garden looking at Jupiter or

Heidi Hammel at an observatory in Hawai'i pioneering the study of Uranus and Neptune, Kepler and TESS and their earthbound predecessors, from October 1995 to July 12, 2022, embodied the discovery stage of the search for exoplanets.

Not that Webb wouldn't be in the discovery stage too. On September 1, 2022, less than two months into its science mission, Webb announced its first *direct* detection of an exoplanet, HIP 65426 b — undoubtedly the first of many to come. Two of Webb's cameras, the Near-Infrared Camera and the Mid-Infrared Instrument, were equipped with coronagraphs to block the light from a host star in order to isolate an orbiting planet. Hubble, too, had made its share of direct observations of exoplanets. But in order to be visible to Hubble, such exoplanets had to be at a great distance from the star. HIP 65426 b also was at a great distance from its host star — one hundred times as far from its star as the Earth is from the Sun. For Webb, that detection had been notable as a first, but as was the case with Hubble's detections of exoplanets, it didn't add much to the search for signs of life.

Unlike Hubble, however, Webb *was* in the business of detecting planets in the habitable zone. To do so, it too would use the transit method. But it also would then deploy spectroscopy, thereby revealing what was in an exoplanet's atmosphere and sometimes even the composition of the planet itself.

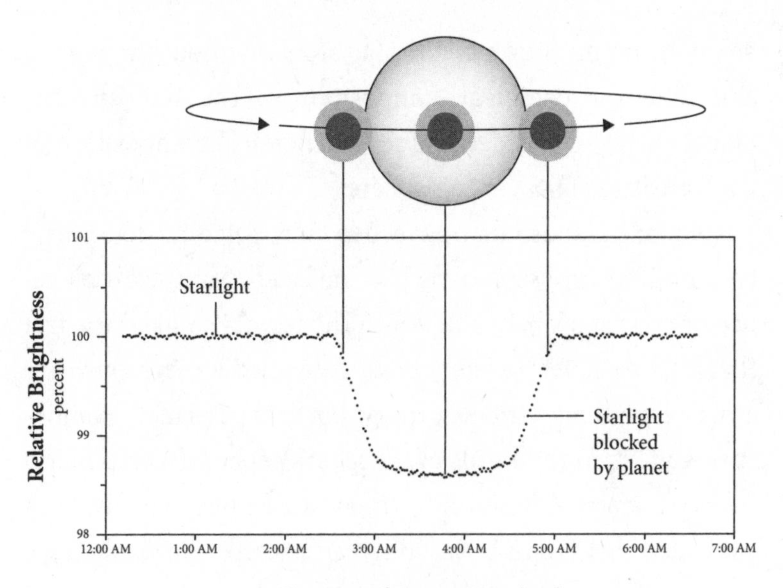

In transit: When a planet passes in front of a star (from the perspective of the telescope), it blocks some of the light. This illustration, along with the actual light curve from Webb's Near-Infrared Imager and Slitless Spectrograph, shows the change in brightness of light from the WASP-96 star system, and vividly displays one such transit. Astronomers can use this data to determine a planet's size and orbital period as well as other essential information, but by comparing spectra from the star alone and from the star and planet combined, they can also discern the contents of the planet's atmosphere.

The Webb process worked like this: First, astronomers would take a spectroscopic reading of the light from a star alone. Then they would take readings as an exoplanet transited the star. By comparing the two sets of readings — the chemical composition of the star alone, without the planet, and then the chemical composition of the star *and* the planet—they could determine the chemical composition of the transiting object's atmosphere.

Nikku Madhusudhan's Webb team targeted K2-18 b partly because Kepler observations had indicated that the exoplanet occupies its star's habitable zone. The team also liked the fact that the exoplanet was larger and contained a more extensive atmosphere than a smaller, rocky, Earth-type planet, qualities that made it, in the words of the team's paper in *Astrophysical Journal Letters,* "significantly more accessible." Finally, they suspected that K2-18 b was a hycean planet—*hycean* being a portmanteau of the words *hydrogen* (in the atmosphere) and *ocean* (on the surface). At the time of Madhusudhan's team's observation, hycean planets were still hypothetical—the term and the concept date only to 2021. But if hycean planets did exist, not only would K2-18 b be one but, as Madhusudhan liked to say, it would be the "poster child."

Hycean theorists had predicted that if a planet had a hydrogen-rich atmosphere and a surface of water (in whatever state—liquid, solid, gas), then its atmosphere would also contain methane and carbon dioxide. Madhusudhan's collaboration identified both. Those detections were, respectively, at the 5-sigma and 3-sigma confidence levels—*sigma* being scientific shorthand for statistical certainty. The gold standard is 5 sigma, which corresponds to a 99.9999997 percent confidence level, or a 1 in 3,500,000 chance that the result is a coincidence or a fluke. A 3-sigma confidence level is not as "robust," in scientific jargon; it corresponds to approximately a 1 in 740 chance of a coincidence or a fluke. That level of statistical confidence—nearly a 739/740 chance that the result is correct—might sound meager in comparison to a 3,499,999/3,500,000 chance, but it's still at the 99.73 percent confidence level. In short, Madhusudhan's team had detected in K2-18 b's atmosphere the all-but-certain presence of methane and the likely presence of carbon dioxide.

These findings were important in themselves. But the detection of methane had a separate significance in the field of exoplanet astronomy. Despite theorists' predictions of its presence in hycean (or at least hycean-like) atmospheres, pre-Webb spectroscopic observations of exoplanet atmospheres had failed to detect any methane. That absence had even earned a nickname: "the missing-methane problem." Madhusudhan's team was not the first to report the identification of methane in an exoplanet's atmosphere. But their discovery provided reassurance to the community that the methane was missing no more.

As for the "detection" of dimethyl sulfide: The paper in *Astrophysical Journal Letters* did offer a sigma range, but even the high end was too low for scientists to regard the possibility of its existence with little more than curiosity and maybe a mental Post-it. Still, the paper couldn't *not* mention it, since most of this molecule in the Earth's atmosphere originates in phytoplankton—microalgae in marine environments. Not that they're animal life. They're not. They're plant life (*phyto* is Greek for "plant"). *Microscopic* plant life at that.

But life all the same.

So the authors of the paper devoted an appropriate amount of space to weighing the evidence for the presence of dimethyl sulfide on K2-18 b, in part so as to preserve any future need to claim priority for a discovery of "I kid you not" momentousness. But they also devoted an appropriate amount of space to clarifying the tentative nature of the discovery. In the body of the paper they characterized the evidence as "marginal" and as being the result of "potential inference." In the abstract of the paper they deployed a phrase that was triply cautious: "suggests potential signs."

Suggests. Meaning maybe.

Potential. Meaning maybe.

Signs. Meaning... what?

Let's pause here and, for just a few pages, take our leave of not just Webb's investigations of extrasolar planets but the entirety of the telescope's tour of the universe. Let's instead consider a question that Webb scientists get all the time. It's a question that applies to all of Webb's goals—to each realm that lies past every fresh horizon. The question is: *Is this what I would see?*

Is this (whatever the *this* is) what I or anyone else would see if we were nearby, in space, eyeing it for ourselves?

The short answer is *No.* Webb sees in the infrared; our eyes don't.

So what *do* astronomers see? What are the rest of us seeing?

In the case of the abstract of Madhusudhan's paper, "signs" referred to spectroscopic analysis—a graph in the form of spikes and troughs, similar to the tracings on a lie detector or an electrocardiogram. Those peaks and valleys indicate the concentrations of or the absorptions of the elements or compounds corresponding to those precise slices of the electromagnetic spectrum—the spectroscopic lines that Fraunhofer discovered in 1815 and Bunsen and Kirchhoff identified several decades later. While our eyes see electromagnetic waves with lengths that fall between 0.4 and 0.7 microns—a span of only 0.3 microns— Webb's infrared range spans 27.9 microns, or more than ninety times the range of human eyesight. Webb's reach into the infrared allows for a greater range of spectral lines and therefore a broader array of chemical identifications than were available to most previous telescopes.

But Webb's two spectrometers also offer greater precision: Webb's spectral graphs contain error bars that are a small fraction of earlier generations' error bars. To an astronomer, a Webb graph could be a thing of beauty, a museum-worthy work of art. At their first viewing of a Webb spectrum, astronomers didn't even have to know what they were looking at to know that it was extraordinary. When presenters at astronomy meetings showed Webb plots, the response was often a roomful of gasps, even if most members of the audience didn't share the presenter's area of specialization. They could transpose the plot to their own fields of expertise, and they could understand what that level of specificity might mean, and they thought—well, as one astronomer said of a graph of her own, confident that she'd get a knowing laugh from the crowd, which she did: "But that phase curve"—a spectroscopic plot that follows an exoplanet not during a transit but during the entirety of its orbit—"*damn.*"

To a nonscientist, however, a plot of spectra might be just...a plot. To a mathophobic nonscientist, it might be worse: a riot of seismograph-like tracings and a pair of axes and a passel of identifiers. There's a reason the initial public release of five images on July 12, 2022, included only one spectrum: The public presumably wouldn't fully appreciate what it was seeing. But everyone, scientists and nonscientists alike, could marvel at the whirligig majesty of a spiral galaxy.

Webb's photographs originate in the same form as the images on any digital device, including your smartphone: as a series of zeros and ones. Software then resolves the numbers into pixels. Each pixel has one of 65,536 possible values, each corresponding to the intensity of the source photons and thus to a specific shade of gray.

The color comes next. Not that the images are in color. They're still in grayscale. But in proposing a program on Webb, astronomers choose from among twenty-nine filters on Webb's imaging instruments that allow them to observe at wavelengths corresponding to whatever elements and compounds they hope to study, and only *then*—after the digital data from those filters has passed through the Webb pipeline—do the image specialists and astronomers on Earth assign colors to the data at those wavelengths. What colors they assign, however, are up to them.

Astronomy (to the bafflement of even many astronomers) doesn't have a uniform standard for assigning colors. Some astronomers in some situations prefer to use deeper and deeper shades of red to designate hotter and hotter temperatures. Others prefer to use deeper and deeper shades of blue for the same designation. Hubble's 2018 version of the Pillars of Creation, for example, assigned hydrogen the color green; Webb's 2022 version assigned hydrogen the color blue. In the case of the "hourglass" ballooning on either side of the protostar at the heart of dark cloud L1527, the big blue globule was where the dust was thinnest, and the big orange globule was where the dust was thickest. If an astronomer didn't have the decoder ring to solve the cipher in a specific image, the message would be meaningless.*

---

* The astronomers and imaging specialists can also make aesthetic choices having to do with orientation, cropping, and contrast. The scientists themselves—like Heidi Hammel misting up at an image of Neptune's rings on a home computer—are just as susceptible as the public to the synapse-snapping allure of a pretty picture. At one astronomy conference, a particularly dramatic Webb image drew a collective gasp.

"That's not science," the presenter said, drily.

The public, of course, doesn't have that decoder ring, but when Webb releases a photo, the description will include, if necessary, a reference to what the colors mean.

Still, you can try this at home:

The images from various space missions, including Webb, reside in the Barbara A. Mikulski Archive for Space Telescopes, in honor of the former senator who saved Webb. Nearly all of the Webb images are available immediately upon download—not just to scientists but to anyone. (The rest are under a brief embargo due to preexisting research restrictions.) The data is the public's to contemplate, to marvel at, but also to play with.*

And play with it the public does. The internet is full of Webb images that are aesthetically inventive. You can create your own Pillars of Creation. They will be useless as science, but they will be uniquely your own. They will be your versions of what these images "really" look like.

And then you can tell friends, family, and strangers on the internet that you have created your own universe—and that you kid them not.

* See: https://archive.stsci.edu/missions-and-data/jwst.

CHAPTER FIVE

THIRD HORIZON: ACROSS THE UNIVERSE

"I don't believe it."

The collaboration had just begun discussing their latest find — one that some members had entered the meeting thinking was "potentially major-paper-worthy," in the words of the team leader, Ori Fox — and already they had a dissenter.

Fox slumped in his seat. His postdoc, Melissa Shahbandeh, sitting at the same long table, shifted in hers. Several of the other members of the collaboration were also present in person in a conference room at the Rotunda, a Space Telescope Science Institute annex in a nearby shopping mall, a couple of floors up from Floyd's 99 Barbershop (through which one had to cut to get to the elevator if the entrance fronting the mall's parking lot was locked). Other members of the group were faces on a screen at the far end of the room. Everyone made their own adjustments, whether physical or mental, to their colleague's opinion.

Weren't they supposed to be having the time of their lives? Just a few evenings earlier they'd all met on Slack to download their data as it came through the Webb pipeline. There had been the shouting of *Wows*, then the sharing of screenshots. The data was plentiful. It was full of rich details. Most important in the moment, it was highly promising. Of what, nobody could be sure. Further elation would just have to wait until somebody had coded the data into comprehensibility.

Which is what Fox and, mostly, Shahbandeh had done. And their efforts had yielded the results that they had been expecting to celebrate. Instead, one of their own had expressed his disbelief. But even before Fox and Shahbandeh could try to bring him around to their way of thinking, another team member spoke up: He didn't believe it either.

Then the two contrarian voices joined, reinforcing each other.

No way, they said. *No way there's that much dust*—the target of the team's Webb observations and potentially a clue to one of the universe's key growth mechanisms. Besides, they said, the data had come too easily. It had come too fast. Like, *right away* fast.

No way, they repeated. *Not right off the bat.* Especially since the method they had used was only one of several possible approaches.

Yes, Fox acknowledged, the result was difficult to believe. But that didn't make it wrong. Fox explained that he and Shahbandeh had done their due diligence. They had double-checked their data. They had challenged their theoretical models. They had recruited a colleague who was an *expert* at that—at challenging theoretical models—and he had agreed that the result was robust and worth reporting.

"He's one of the best in the business," Fox said. "That's why he's *on our team.*"

The two holdouts still didn't believe it.

"Oh, come *on,*" Fox said.

But the verdict was in. And the collaboration couldn't go ahead with writing up the result and submitting it to a peer-reviewed journal until the sentiment among the team members was unanimous.

Finally, Fox folded. He addressed the two holdouts.

"Fine," he said. "How much more work do we need to do?"

Before October 6, 1923, the universe was divided in four: Earth, as always; then the sundry components of the rest of the solar system; then the realm of the stars that William Herschel and his colossal mirrors had endowed with the dimension of depth; and then . . . anything else?

Maybe not. Maybe the stars were as far as it goes—*it* being the universe. Maybe after Herschel "broke through the barriers of the heavens," as his epitaph said, telescopes had reached the deepest depths of space.

But why should the farthest stars define the outermost boundary of existence? Every time astronomers had sufficiently extended the limits of telescope technology, they had also extended the limits of the universe they'd inherited. And if something more *was* out there, beyond the farthest horizon, astronomers at the turn of the twentieth century had already identified what it might be: *nebulae*, from the Latin for clouds.

Not that they knew what these clouds were, other than faint smudges in the night sky. Before Galileo, naked-eye observers had counted nine such smudges. Galileo, thanks to his telescopes' powers of magnification, had discerned that some of those clouds were simply "congeries"—jumbles—"of innumerable stars." The identity of the other clouds, though, remained elusive, even as astronomers discovered more and more of them. By the time Herschel began observing stars in the 1770s, the number of nebulae had risen to ninety. Through his telescopes' powers of depth—their ability to collect greater and greater amounts of light, allowing Herschel to see deeper and deeper into the stellar realm—he had identified another twenty-five hundred.

Herschel spent most of his observing time on stars, but he also tried to discern the nature of those nebulae that he couldn't resolve. Were they single stars, each floating in some sort of shell? Were they gaseous matter? Or were they, in a term just then emerging, "island universes" like our own?

Like our own *what*, though? The ancient Greeks had a name for the whitish streak that spills across the heavens on a dark night: *galactos*, or "the milky thing in the sky."* Over the centuries the two terms had become nearly interchangeable—galaxy, Milky Way. After the invention of the telescope, astronomers determined that the milky thing was indeed composed of stars. By Herschel's time, some astronomers and philosophers (including Immanuel Kant, in 1755) had granted the likelihood that the stars in our island universe were other suns like our own, and then they extended the logic: If our Sun was one among a multitude of stars we call the Milky Way, then maybe the Milky Way was one among a multitude of nebulae we call island universes.

In the 1840s—the next generation after Herschel's—William Parsons, Third Earl of Rosse, found inspiration in Herschel's advocacy for mirrors in astronomy and oversaw the construction of a giant reflecting telescope on his family's estate in County Offaly, Ireland. The structure quickly earned the nickname the Leviathan of Parsonstown. The diameter of its mirror was 72 inches; Herschel's largest had been 48 inches—a difference of two feet. The pattern of

* Other cultures had their own names. In China, for instance, it was the Silver River.

discovery held: More light meant more depth; more depth meant more nebulae.

But more light also meant higher resolution—a greater delineation of detail. Rosse (as Parsons was known) found not only that he could look deeper into space and thereby discover more nebulae but that in some cases he could identify the essence of a nebula's structure. In 1850 he published a paper that included drawings of five nebulae in the shape of a spiral as well as a list of fourteen more spiral nebulae. The contemporaneous American astronomer Stephen Alexander had already been consulting Herschel's attempts to map "the construction of the heavens" in three dimensions, but now, in 1852, having read Rosse's paper, he had the advantage of knowing what shape to look for. He soon concluded that the stars in our island universe—our galaxy, our Milky Way—also "constitute a spiral."

Well into the twentieth century, the question of whether the spiral nebulae resided inside or outside our galaxy remained a matter of debate—indeed, a "Great Debate," as historians would come to call it. In 1920 two of the most influential astronomers of the day parried on the stage of stately Baird Auditorium at the Smithsonian Museum of Natural History, on the National Mall in Washington, DC. Forensics dexterity, however, wasn't going to settle the argument. Only evidence was.

It came three years later. On October 4, 1923, at the Mount Wilson Observatory in the San Gabriel Mountains northeast of Los Angeles, Edwin Hubble focused on the Great Andromeda Nebula, or M31.* The telescope he was using, only four years old, was anchored by a 100-inch mirror—a 28-inch advance over Rosse's 24-inch advance† over Herschel's 48-inch behemoth. But Hubble had another advantage over those big-mirror predecessors: photography.

This relatively new technology offered two advantages to astronomers. The first was the amount of light. Both reflecting mirrors and refracting lenses accumulate light, but the observation of that accumulation of light happens only between blinks of an eye. The accumulation of light on a photographic plate also happens between blinks of an eye, but in photography that eye is a shutter, and that shutter can remain open for as long as the photographer chooses. All the while the glass plate will be gathering light and then more light, the photons from distant phenomena saturating the silver bromide. On this occasion, for the purpose of photographing M31, Hubble kept his shutter open for forty-five minutes.

The next day Hubble exploited the second major advantage of photography. An eye might capture this or that impression of an image. The brain might notice this or that

---

* In the classification system of 103 nebulae that the French astronomer Charles Messier published in 1781. To Messier, the nebulae were nuisances that impeded his search for comets. For Herschel, Messier's *Catalogue* was an invaluable introduction to the study of nebulae.

† However immediately useful, the Leviathan was ultimately disappointing. Not surprisingly, the wet, windswept conditions common in Ireland frequently led to what astronomers call "bad seeing."

detail within the image. A pencil on a pad might attempt to approximate this or that detail within an image.

All of those transactions were highly subjective.

The image on a photographic plate, however, could be considered impartial, an image that was relatively free of the original observer's bias. (It was, of course, still subject to the interpretations and biases of the follow-up investigator, if only because science is a human endeavor.) Now that image was permanent; it was part of the historical record. Astronomers could return to it at any time and study it at their leisure.

Which was what Hubble did in his office the following day: examine the plate from the previous evening. As he did so, he thought he noted among the spiral arms of M31 a "nova," or new star (actually, as astronomers were then beginning to realize, an old star that has exploded). So Hubble returned to M31 that night and took another forty-five-minute exposure. The following day he gathered the plates from the previous two evenings and began comparing them with earlier photographs of M31 from a number of different dates.

What he'd discovered, he realized, wasn't a nova. It wasn't a star that brightens suddenly, then dims, then disappears. It was, instead, a variable — a kind of star that, as its name suggests, varies. It pulsates. It brightens, then dims; brightens, then dims; brightens, then dims. Hubble could sequence the plates like frames in a motion picture — that newfangled source of mass entertainment — and witness the brightening and dimming for himself.

Upon this closer inspection, Hubble realized that the variable in M31 wasn't just a variable. It was a Cepheid variable—a version that doesn't just brighten and dim, brighten and dim, but does so with (as the English amateur astronomer John Goodricke, in 1784, had first identified in the star Delta Cephei, in the constellation Cepheus) clockwork regularity and consistent intensity.

A Cepheid variable was what Hubble had been hoping to find in his quest to determine whether the M31 nebula was part of our galaxy or beyond it. In 1908 the Harvard astronomer Henrietta Swan Leavitt had discovered a proportional relationship between the pulsation period of a Cepheid variable and its intrinsic brightness: the longer the period from peak to peak, the brighter the variable. Cepheid variables were, in short, what astronomers call a "standard candle"—a class of objects whose luminosity doesn't change. A 100-watt light bulb, for instance, is a standard candle: If you know that its absolute magnitude is 100 watts, then you can apply the inverse-square law to its apparent magnitude—how bright it looks to you at whatever distance—and calculate how far away it actually is. When Hubble compared the pulsation period of the Cepheid he'd found in M31 with the pulsation periods of the Cepheids that Leavitt had collected, he concluded that it was far enough away that it (and therefore its host nebula, M31) lay beyond our own galaxy of stars.

Hubble went back to H335H, the photographic plate he'd

made on the night of October 5–6, crossed out the "N" he'd used to designate "nova," and added a celebratory "VAR!"*

That same year the American astronomer Vesto Slipher published a list of forty-one nebulae, most of which exhibited a curious trait. Through a combination of photography and spectroscopy he had found that the spectral lines in thirty-six of those nebulae didn't align with the positions on the electromagnetic spectrum where one might expect to find them. Instead, they occupied positions toward the red end of the visible portion of the electromagnetic spectrum. Those "redshifts" indicated that the wavelengths had stretched, which is what you would expect if the object emitting the light was moving away from you (or if you were moving away from it). The larger the shift, the greater the recessional velocity.

Over the next few years Hubble began plotting on an *x*/*y* graph the key data of those nebulae and also of a few more that a Mount Wilson colleague collected. He placed the distances of the nebulae (which he inferred from Cepheid variables as well as, less reliably, the nebulae's relative luminosities) on the *x*-axis and the velocities (which he inferred from their redshifts) on the *y*-axis. The plot suggested† a proportional relationship: the more distant the nebula, the greater the

* I was once in the office of Hubble's protégé Allan Sandage, who asked if I wanted to see the plate. He reached into a file cabinet (!), produced it, and placed it in my hands. "In red, he wrote it," Sandage said of Hubble's "VAR!" inscription. "He knew reporters were going to come and look at this."

† *Suggested* because the error bars wouldn't pass peer review today.

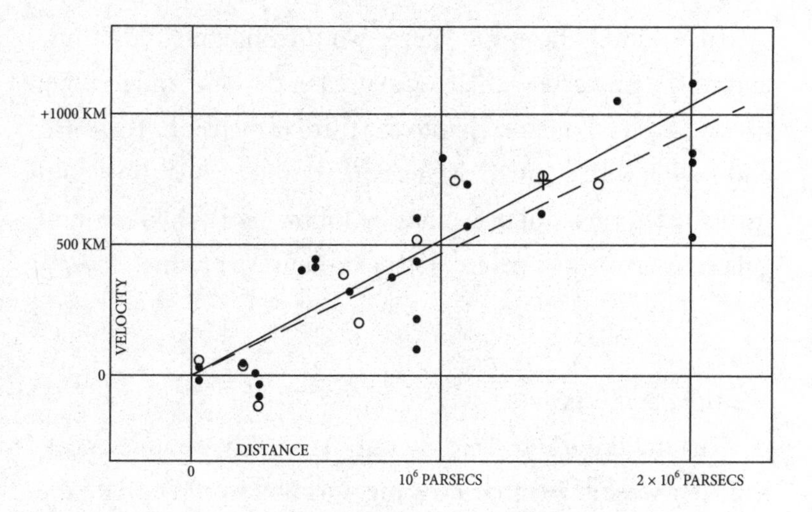

The fourth dimension: In 1929 astronomer Edwin Hubble published a graph showing a direct relationship between the distance of galaxies from us and the velocities at which they are receding from us (and us from them). Here was the new science of cosmology in fetal form: a universe in motion over time.

---

redshift—the farther, the faster. Hubble concluded—as had, independently, the Belgian theorist Georges Lemaître, working not from empirical data but from Einstein's equations for general relativity, which allowed for changes in the geometry of space—that the universe is expanding.

In breaking through "the barriers of the heavens," William Herschel had endowed the night sky beyond our solar system with the third dimension of depth. Hubble and Lemaître extended that dimension, and with it the size of the universe, beyond our galaxy.

In so doing, however, these two astronomers didn't just endow the universe with an even greater depth than Herschel had, an even farther reach into the third dimension. They also endowed it with a fourth dimension, the one that Webb would investigate in a similar fashion by following the evolution of galaxies across the universe — the dimension of time.

We still don't believe it.

The holdouts kept holding out. At weekly webinar meeting after weekly webinar meeting, Ori Fox would update the team, and the answer from the holdouts was always the same, as was the reasoning:

Observation and theory are mutually dependent. If you're trying to match, or "fit," empirical data to a theoretical model, each needs to be of sufficient quality to meet the needs of the other. In the "old days" — pre-Webb — astronomers didn't have access to data of such high quality. They could hardly imagine data of such high quality. But now that they did have access to that quality of data, they needed models that met an equally high standard — models that were equally pristine.

Fox and Shahbandeh didn't disagree with this reasoning. How could they? The collaboration was using an instrument nobody had ever used before, and they were seeing things nobody had ever seen before. More to the point, though, they were using that instrument to address an anomaly of nature that had long puzzled astronomers:

Dust, dust, everywhere.

The universe had too much of it. Just as Webb astronomers studying our solar system and our galaxy looked to water for clues regarding star and planetary formations (as well as for the potential for life), so did many Webb experts studying the evolution of galaxies prioritize dust. Put enough dust together and it will gather gravitationally into clumps. Over time those clumps of dust will interact gravitationally with other clumps of dust, the collective clumps growing in both size and density. Eventually those clumps will compactify into concrete objects on scales from micrometeoroids to galaxies. And whatever dust is left over will occupy nearby space, whether interstellar (between stars within a galaxy) or intergalactic (between galaxies).

But for half a century or so, astronomers had been finding more dust than they expected. Because the speed of light is finite, astronomers studying objects at greater and greater distances across the universe were looking farther and farther back in time. That was what they wanted to do, since they were trying to trace growth patterns from the universe's infancy to the present day. But cosmologists studying volumes of dust at specific times in the universe's history had consistently found more dust than their models predicted.

Something had to be the source of all that dust, and astronomers had converged on a likely suspect: supernovae—massive stars that end their lives in explosions (or, more accurately, implosions that collapse the progenitor stars' cores, triggering nuclear reactions that culminate in explosions). Until Webb, though, no telescope had been able to observe the evidence at the right wavelengths.

Stellar dust is usually cold, at least relative to the heat of stars, and cold dust peaks at about 20 microns on the electromagnetic spectrum. The Spitzer Space Telescope, active from 2003 to 2020, could observe at 20 microns, but it wasn't powerful enough to examine even relatively nearby supernovae to any meaningful extent. Then, in 2010, the Atacama Large Millimeter Array in Chile and the Herschel Space Observatory turned their infrared detectors to Supernova 1987A,* a star that in 1987 exploded in the Large Magellanic Cloud, a satellite galaxy of our own Milky Way—right next door, cosmologically speaking (to be precise, 158,200 light-years away). Those observations revealed that the remnants of the supernova were still awash in the dust it had ejected and that the amount of dust was impressive. In fact, it was so voluminous that if Supernova 1987A was typical of supernovae throughout the universe, then supernovae might indeed account for the seeming overabundance of dust.

The strategy that Fox's team chose was to explore other supernovae at the possible cold-dust peak of 20 microns. Those supernovae would be relatively nearby (from ten to one hundred times the distance to the Large Magellanic Cloud)

---

* The designations of supernovae follow a simple pattern: year, then chronological order by alphabet. Supernova 1987A was therefore the first supernova discovered in 1987.

but beyond the reach of the Atacama Array and Herschel. Were the dust remnants of those supernovae consistent with Supernova 1987A's? If so, that consistency would lend further support to the idea that supernovae had seeded the universe with the dust that formed the basis of, well, everything.

Using Webb's Mid-Infrared Instrument, Fox and his collaborators could place one photographic filter at the 20-micron point corresponding to a possible cold-dust sweet spot and another at the 5-micron point corresponding to the hotter emissions of a supernova's nearby stars. The first among their initial five targets was a supernova within NGC 6946,* a galaxy 22 million light-years away from Earth that astronomers had nicknamed the Fireworks Galaxy because of its relatively prodigious output of supernovae (about a dozen in the last century).

Fox's team had only a rough idea as to where in the Fireworks Galaxy their supernova might reside. They knew that, like all astronomers using the Mid-Infrared Instrument, they didn't *need* to convert the strings of zeros and ones into a photograph. They could create an algorithm that would search the binary data for telltale pixels.

But in an image, Fox and Shahbandeh felt, the supernova might just...*pop*. It might just be evident at a glance: Everywhere would be the blue stars—because blue was the color they had chosen to assign to the filter at the lower end of the

---

* NGC is an abbreviation of *New General Catalogue of Nebulae and Clusters of Stars,* a Royal Astronomical Society publication from 1888 that listed 7,840 objects.

spectrum, where the light waves are shorter and therefore more energetic and therefore hotter. And somewhere among the assemblage of blue stars would be the red residue of the supernova—because red was the color they had chosen to assign to the filter at the higher end of the spectrum, where the light waves are longer and therefore less energetic and therefore cooler.

So even as the deluge of first data from the observations of NGC 6946 burst forth from the Webb pipeline, even before convening the first webinar meeting about Supernova 2004et* at the Rotunda annex of the Space Telescope Science Institute, Fox and Shahbandeh had commissioned images.

At that point in the process the images were still black and white, showing only the brightness or dimness of the individual pixels that Webb technicians had derived from the initial zeros and ones. Fox and Shahbandeh looked at an image from the 5-micron filter first. It was resplendent with points of light. Stars and stars and stars—just as the collaboration had expected when choosing to observe at the 5-micron level.

---

* So proficient did the detection of supernovae become after 1987A—the supernova that basically inspired a new era of supernova science—that astronomers soon needed to revisit the rules of their nomenclature. Once they ran through the A-to-Z designations in chronological order of detection over the course of a year, they circled back to the beginning, this time lowercasing the initial letter (from A to a) before running through the alphabet again (aa, ab, and so on, through ax, ay, az). When they exhausted *that* list, they moved on to the next initial letter (ba, bb, bc, . . . , ca, cb, cc, and so on). In short: The designation 2004et means it was the 150th supernova that year.

Then Fox and Shahbandeh looked at an image from the 20-micron filter. All of those pinpoints from the 5-micron dropped out — just vanished. All except one.

That pinpoint of light not only didn't vanish, it intensified.

Fox hovered. Over Shahbandeh's shoulder he asked about scaling the image and optimizing the content and changing the contrast. He didn't *not* know what he was talking about, but after a while he recognized the strategic advantage of being quiet and deferring to the resident expert.

Shahbandeh finally ran the code colorizing the filtered images. Then she stacked them into one composite image.

When that image appeared on her computer screen, she and Fox both reared back. The image was a field of blue dots. They were *all* blue — all except one. It was red. It was cool. It was debris, ejecta, dust.

It was Supernova 2004et.

It *popped,* all right.

The idea that supernovae are seeding the universe wasn't new. The search for their contributions to interstellar dust was relatively recent — and thanks to Webb, that search was entering the next era. But scientists had known for decades that the dust particles from supernovae are the engines of evolution, both cosmic and human.*

---

* *Evolution* is a word that NASA had avoided since at least the first half-decade of the twenty-first century, so as not to displease creationists.

It popped: Ori Fox and Melissa Shahbandeh could have extracted the data they needed from digital algorithms without resorting to images, but they wanted to see for themselves the extent of the contrast between Supernova 2004et and the other stars in its host galaxy. (This preliminary image is the one that stunned them on a visceral level. A higher-resolution version would appear in the peer-reviewed paper.)

Galaxies and the stars within them begin their lives in the interstellar medium. That medium is full of molecular

Hence the official designation of this goal among Webb's four official goals as "Assembly of Galaxies," even though astronomers call the concept "galactic evolution."

gas and dust. That gas and dust coalesce into clouds, and the merging of clouds creates the conditions for the birth of a star. The star blazes. It burns. It might host planets, and it might even provide those planets with the nutrients necessary for life. Eventually the star dies. If it's not relatively all that massive, like the Sun, it slowly decays. If it's massive enough, though, it goes supernova, ejecting its contents into the interstellar medium.

Then the cycle repeats: interstellar medium; clouds of molecular gas; collisions of clouds; births of stars; deaths of stars, some in the form of supernovae, which seed the interstellar medium.

And the cycle repeats. And repeats and repeats and repeats over millions and billions of years. And with each repetition in the cycle, the universe would get a little more complex, as physicists had known since the mid-1950s.

Over an eighteen-month period, four physicists at the Kellogg Radiation Laboratory at Caltech had worked in a windowless room, scribbling on a blackboard, feverishly trying to figure out the origins of the elements. When a star goes supernova and ejects its contents back into the universe whence they came, might the thermonuclear explosion add a little extra? Can the chain of reactions causing the star to explode also rip apart the basic building blocks of matter and compress them back together again to create new and heavier elements? Could successive generations of supernovae account for the elements in the periodic table?

They could, the four physicists concluded.

The result of the Caltech collaboration, in 1957, was a 104-page paper in the journal *Reviews of Modern Physics* titled "Synthesis of the Elements in Stars." The four authors* had done for the origin of elements what Darwin had done nearly a century earlier for the origin of species—and they wanted the world to know it. Echoing the final line from Darwin's *Origin of Species*—"from so simple a beginning endless forms most beautiful and most wonderful have been, and are being, evolved"—they wrote, "The elements have evolved, and are evolving." Or, as Joni Mitchell put it in her song "Woodstock" a decade later, "We are stardust."

As soon as Webb entered scientific mode in July 2022, an international collaboration of more than one hundred researchers—"Physics at High Angular resolution in Nearby GalaxieS," or PHANGS (at least in the somewhat unorthodox acronym the scientists themselves had styled)—began targeting spiral galaxies. This species of galaxy (our Milky Way is one such specimen) looks like pinwheels, which is what they are. They're spinning, though so slowly from our perspective that we can't see the rotation with the naked eye.† Still, on a galaxy's own scale of cosmic time, the rotational rate is sufficient to herd the gas and dust and stars into the

* E. Margaret Burbidge, G. R. Burbidge, William A. Fowler, and F. Hoyle, or B^2FH, as scientists call the collaboration to this day.

† Astronomers can nonetheless determine a rate of rotation by measuring the redshift in the "arms" of a disk as it rotates away from us and the blueshift in the arms as the disk rotates toward us.

signature arms within the spiral. Each galaxy in the PHANGS observing program contained hundreds of billions of stars, but the arms had heretofore remained at least somewhat opaque because of the omnipresence of dust and interstellar gas.

In the years leading up to Webb's launch, PHANGS had targeted nineteen spiral galaxies in optical, radio, and ultraviolet regimes using Hubble as well as telescopes on Earth. But to pierce the spiral arms they would need infrared—which meant they would need Webb.

In part the purpose of PHANGS was to pierce the spiral arms in order to reveal the *components* within. They used Webb's Mid-Infrared Instrument to study galaxies at 7.7 and 11.3 microns and Webb's Near-Infrared Camera to study galaxies at 3.3 microns—wavelengths that correspond to molecules and hydrocarbons that play a crucial role in star and planet formations. Webb detected them at once.

In part the purpose of PHANGS was to pierce the spiral arms in order to reveal the *structures* within. Astronomers not only found enormous bubbles of gas and cavities of dust but also observed how the bubbles and cavities interact.

And in part the purpose of PHANGS was to pierce the spiral arms in order to begin to piece together the *evolutionary process* in detail: the rate and timescales of star formation, and how one generation of star formation can feed (or, if it's violent enough, impede) the next generation of star formation.

But the overall purpose of PHANGS wasn't just to advance

the science of the collaboration members. It was also to advance the science of the community. PHANGS was producing a "treasury survey," a trove of research that anyone could access, a resource for future observations and adjustments to theoretical models.* Within the first six months of Webb operations the collaboration had studied only four of the eventual nineteen target galaxies, but such were the quality and quantity of data that even that preliminary research filled an entire issue of *Astrophysical Journal Letters*—twenty-one papers in all.

Webb wasn't all-seeing in the infrared. It still couldn't cut through the densest clouds of dust and gas. For instance, on July 16, 2022, only five days into its scientific program, Webb's Near-Infrared Camera observed the remnants of Supernova 1987A and failed to image the neutron star that, according to the laws of physics, should be what remains of the post-supernova star. What Webb could do, though, was identify the presence of fluorescing argon and sulfur that, as a team member of that collaboration said, was the "smoking gun" for the presence of a neutron star.

* Webb is good at that—creating vast inventories for future research. When a team led by Svea Hernandez, a European Space Agency and AURA astronomer at the Space Telescope Science Institute and one of the foremost experts in galactic evolution, submitted a study of one spiral galaxy, the referee told her, only half jokingly, *Just supply the data, and that's your paper.*

We believe it.

The holdouts finally gave their blessing. Three months had passed since the data download, and the initial excitement was long gone, at least for Fox. But finally his and Shahbandeh's rechecking of data and rechallenging of models had satisfied their two colleagues. The former holdouts were willing to put their names on a paper reporting the results.

On January 25, 2023, the team submitted that paper to the *Monthly Notices of the Royal Astronomical Society*, with Shahbandeh as the lead author. By then they had reduced the data on another supernova in the Fireworks Galaxy, 2017eaw. They had taken the amounts of dust present in the remnants of those two supernovae today and compared them to the amounts of dust from 1987A when it was their same ages. While the two new supernovae hadn't exhibited the same level of dust as 1987A, they'd come close—close enough, anyway, that the team could claim in the paper's abstract that their results supported the hypothesis that supernovae just might account for the otherwise inexplicable amount of dust in the universe.

As with Webb's studies of water in our immediate or at least immediate-ish vicinity—in the solar system and within our own galaxy—so Webb's studies of dust in galaxy after galaxy after galaxy spanning billions of light-years were, ultimately, a contemplation of our origins, on both a metaphorical and a literal level.

But Webb still had one more horizon to cross before it could reach its ultimate destination: the epoch of stars and supernovae and galaxies first coming to light—the dawn of the elements that would one day synthesize into the stuff of us.

CHAPTER SIX

FINAL HORIZON: IN THE BEGINNING

Somewhere in there was a signal. Rebecca Larson was sure.

As sure as she'd been with her colleagues earlier that evening, while they were all sitting around the coffee table in the hotel lobby in Seattle staring forlornly at a laptop. As sure as she'd been with her supervisor, Dan Coe, throughout the opening reception of the American Astronomical Society meeting and then over dinner, whenever he'd whispered to ask whether she really, truly thought she saw a signal among the noise. Her answer had never wavered.

I'm sure.

But finally the evening had ended, and she and her colleagues had dispersed—they presumably to their beds to sleep, and she to hers to code. To open her laptop and get to work. To check the data. To challenge the model. To challenge the pipeline itself—the master code through which Webb's zeros and ones passed between download from the telescope and delivery to the scientists. To account, even, for the fact that Webb could wobble a bit during observations.

Webb was still a novelty. Astronomers were still adapting to its idiosyncrasies. Larson had learned spectroscopy during graduate school while working on one of the Keck telescopes in Hawai'i—and *that* telescope's nuances she understood intimately. But Webb's? She was learning on the job, just like everyone else.

The usual questions guided her as she clicked away on her laptop in her hotel room. How far into the infrared had the spectrum shifted—the redshift that would indicate the age of the object? How many emission lines were showing up—the data that would reveal the chemical composition of the target? Could she match even one faint emission peak with a telltale wavelength—a match that would allow her to align all the other faint emission peaks with their familiar locations along the electromagnetic spectrum?

Could she make the early universe snap into place?

Yes.

Somewhere in the vicinity of 3 a.m. Larson texted to Coe a crude version of the signals she'd wrested from the noise.

The plot wasn't beautiful. It wasn't up to her standards. But it was good enough for now. It was everything her collaboration had wanted.

The target was as old as the team had hoped. Its emission lines indicated the presence of the elements they would have expected from a galaxy at such an early stage in the history of the universe. Those emission lines were the most distant—and therefore the earliest—evidence of elements in the universe.

And as a bonus, nobody could accuse them of breaking cosmology.

JAMES WEBB JUST BROKE COSMOLOGY.

HOW THE JAMES WEBB SPACE TELESCOPE BROKE THE UNIVERSE.

HOW JAMES WEBB BROKE COSMOLOGY IN ONLY TWO MONTHS!

The headlines in late 2022 and early 2023 were borderline apocalyptic. *Borderline* because what they portended wasn't quite the end of the universe; it was instead the end of our understanding of the universe. Then again, the universe as we understand it is the only universe we know, so maybe not so borderline after all.

That understanding of the universe was less than a century old. As a cosmologist once said at a Space Telescope Science Institute conference, "A poor soldier who died in the trenches in 1914 knew as much about the universe as a caveman." The cosmologist was being hyperbolic. The caveman wouldn't have known some of what the infantryman might have known: for instance, that some of those moving "stars" up there in the nighttime sky were variations on Earth, or that our home orbits the big bright thing in the daytime sky, not the other way around. But the cosmologist had a point: Our understanding of the universe had advanced more in the past hundred years than in the entire previous history of civilization. That cosmologist's own field—cosmology—hadn't even existed as a science when he was born, in the mid-1950s.

The modern version of cosmology had begun to emerge only when Edwin Hubble and Georges Lemaître, in the late 1920s, presented evidence that the universe was somehow expanding. That realization helped solve a mystery that had haunted astronomy at least since Isaac Newton's introduction of a universal law of gravitation in his *Philosophiae Naturalis Principia Mathematica,* in 1687: If the universe was full of matter that was attracting all the other matter in the universe through gravity, why hadn't the universe collapsed? Six years later, Newton acknowledged to an inquiring cleric that positing a universe full of matter that was in perpetual equilibrium was akin to making "not one Needle only, but an infinite number of them (so many as there are particles in an infinite Space) stand accurately poised upon their Points. Yet I grant it possible," he immediately added, "at least by a divine Power."

"It was a great missed opportunity for theoretical physics," Stephen Hawking wrote in a 1999 introduction to a new translation of the *Principia*. "Newton could have predicted the expansion of the universe."

So, too, Einstein. When, in 1917, he applied his equations for general relativity to the universe as a whole, he confronted the same problem as Newton. Their understandings of gravitation were different, but Einstein's universe was still full of matter influencing all the other matter in the universe, and it hadn't collapsed, either. Unlike Newton, though, Einstein added to the equation not a divine power but the Greek letter lambda (Λ), an arbitrary mathematical shorthand for whatever was keeping the universe in perfect balance.

The expanding universe of Hubble and Lemaître rendered Newton's divine intervention and Einstein's lambda irrelevant (though not necessarily nonexistent). But an obvious question immediately arose: Expanding from what?

Reverse an outward expansion of the universe and you eventually wind up at a starting point, a birth event of sorts. A few scientists, beginning with Lemaître, suggested a kind of "explosion"—or, less dramatically but more accurately, an *expansion*—of space and time, a phenomenon that later acquired the (initially derogatory) moniker "Big Bang." The idea sounded fantastical, and for several decades, in the absence of empirical evidence, most astronomers simply ignored it—a luxury they could no longer afford after the side-by-side publication of two papers in the July 1965 issue of the *Astrophysical Journal*.

In the first paper, four Princeton University theoretical

physicists calculated a temperature for that hypothetical primordial fireball, then followed that temperature forward in time as the universe expanded and cooled until they arrived at a prediction of the current temperature. In the second paper, two Bell Labs radio astronomers reported the serendipitous measurement of that very temperature (they weren't looking for it and didn't even know what they'd found until the Princeton foursome explained its significance to them) in the form of an all-sky microwave imprint, a relic from the early universe matching the most ambitious prediction in the Big Bang theory.

The match of prediction and detection in the former and latter of these two papers was, if not definitive, at least persuasive enough that most of the astronomy community decided that the seeming absurdity of an expanding universe bursting forth from a single point of something-or-other was, in fact, subject to testing and verification. Just like that, cosmology graduated from metaphysics to physics, from hand-waving to hard science. Over the following decades scientists improved, adjusted, and expanded the Big Bang model — testing it, verifying it; matching prediction to detection — until, by the early years of the new century, a Standard Model of Cosmology had emerged.

The emergence of that model fell within the lifetimes of every person working on Webb at the time of its launch. That model was all anyone knew. And then, less than a year later, it was the model that headlines were claiming Webb had broken.

The afternoon after Rebecca Larson texted Coe with her middle-of-the-night bulletin, various members of the collaboration would be presenting three talks back-to-back-to-back in ten-minute increments. The middle talk was going to involve the galaxy that Larson had been working on, though the talk wouldn't include this latest data—unless the team could justify inserting it, if only in passing.

That morning they sent Larson's plot back to the team member in Copenhagen whose own code had produced mostly noise. He reported back that, yes, Larson's data looked robust. So they got to work. They'd have to make a PowerPoint slide. They'd have to figure out where to insert the slide into the presentation. Tiger Yu-Yang Hsiao, a graduate student at Johns Hopkins who would be delivering the talk, would have to figure out what to say and how much time to devote to it. Two other team members, including Coe, would be making their own presentations, for which they needed to prepare too.

But at some point in the middle of all the commotion, Coe quieted the group and made a suggestion: *Take a breath.* See the moment from a broader perspective, he encouraged them, one encompassing the whole of their careers. Remind themselves that they were participating in a once-in-a-lifetime event. Maybe even, if they were fortunate, twice-.

It was Coe's second, anyway. His first had come on a Saturday afternoon some ten years earlier when, as a PhD candidate at Johns Hopkins, he had been working at his apartment, a ten-minute walk from campus. He was standing at a makeshift desk—a level surface atop some cardboard boxes—studying the Hubble data from a distant galaxy. His collaboration had chosen to observe the galaxy because the Spitzer Space Telescope indicated it was at a high redshift and they wanted to know how high. How far into the red had the expansion of space itself stretched the light from the galaxy—a measure that would allow astronomers to infer the galaxy's age?*

The redshift, Coe's calculations showed, was 10.

But then Coe got to thinking. A redshift of 10 was where he'd set the upper limit for his search. He had assumed that it wouldn't be any higher, but that was only an assumption. What if he removed that restriction? After all, Hubble went to 11.†

Coe soon determined that the redshift of the galaxy was indeed higher than 10. Approaching, but not quite reaching, 11. (Some degree of uncertainty is inevitable.) Which meant that the light had left the galaxy when the universe was somewhere around 400 or 500 million years old. Which meant—

Which meant that, if the data was accurate, Coe had just discovered the most distant object anybody had ever seen.

---

* The formula for using redshift to determine how far light has traveled is complicated. Suffice it to say that a redshift of 0.25 corresponds to a distance of 3 billion light-years; a redshift of 1, to 7.7 billion light-years (or more than halfway back to the Big Bang); a redshift of 10, to 13.2 billion light-years (or about only 600 million years after the Big Bang).

† Insert your own *Spinal Tap* joke here.

His advisor, when Coe reported the result to him, issued the usual cautions: Check the data; challenge the model. Which Coe and the rest of the team did, until they felt confident enough to publish the result. In the end, their galaxy held the redshift record for a little more than three years before ceding to a galaxy with a redshift of 10.957. By then Coe was working full-time at the Space Telescope Science Institute, awaiting the launch of Webb and its superior technology. Even as he accepted a position at the Institute, he knew that he wanted to return to that galaxy, and now he had—a decision that had rewarded Coe with his second once-in-a-lifetime thrill,* his team members with their first, and all of them with surprise after surprise. The galaxy had turned out to be a gift that kept on giving.

Astronomers would have found some significance in the name of the galaxy that Coe had discovered in 2012 and re-observed in 2022: MACS0647-JD. Maybe not so much the "JD" part, which was short for J-band Dropout, a technical term describing the Hubble telescope's alignment of filters

---

* Or maybe his third. In March 2022 he had been part of a collaboration that discovered the most distant single *star,* which they named Earendel, at a redshift of 6.2 (or 12.9 billion light-years). The week before the American Astronomical Society meeting where Coe's team was now gathering, Prince Harry, Duke of Sussex, had published his memoir *Spare*, which included a reference to the star:

"Billions of miles off, and probably long vanished, Earendel is closer to the Big Bang, the moment of Creation, than our own Milky Way, and yet it's somehow still visible to mortal eyes because it's just so awesomely bright and dazzling.

"*That* was my mother."

Coe's wife, drily, to her husband: "You've really made it now."

during the initial observation. The MACS0647 part of the galaxy's name, however, was far more telling. It identified a particular cluster of galaxies in the relatively nearby universe (from our perspective), indicating that the discovery was the result of a method cosmologists call gravitational lensing.

According to Einstein's general theory of relativity, the presence of any object with mass causes a curvature in the confluence of the universe's four dimensions, or space-time. The greater the mass, the greater the curvature. If a foreground galaxy is sufficiently massive, it will deflect the light from an object that, from our point of view, is behind it, and bend that light so that it becomes visible to us. It will also magnify that light and even generate multiple images of the light source. In the case of MACS0647-JD, the foreground galaxy was at a redshift of 0.591, indicating a distance of more than 5 billion light-years, and it provided three images of the background galaxy.

The first surprise for Coe's collaboration came in September 2022, only two months after the telescope began collecting data. Webb's superior level of resolution had determined that MACS0647-JD had *two* components. Maybe MACS0647-JD was a single galaxy, and those components were two stellar complexes within it. But maybe MACS0647-JD was actually two galaxies—in which case those two galaxies might be merging. In that case, MACS0647-JD would be the earliest known galaxy merger yet—only 460 million years after the Big Bang. And maybe it wasn't the only merger the team had captured: Another galaxy seemed to be hovering "nearby," and in *that* case it would likely be merging with MACS0647-JD.

All those results were part of Tiger Hsiao's talk — the one into which he managed to squeeze Larson's spectra.

A little bit underwhelming, Dan Coe thought afterward, regarding the response to Larson's slide. Or lack of response. No evident stirring. No follow-up during the lightning-quick Q&A. The turnout was okay, but, he reminded himself, the talk was in competition with twenty-six other talks throughout the convention hall. And maybe the audience looked at it, saw the emission lines for elements they would expect to see there (the requisite hydrogen, plus carbon, oxygen, neon), and, not seeing any evidence that the Standard Model of Cosmology was broken, moved on, mentally. Also, they had no context for the significance of the slide; no press release had alerted them to await the announcement that the team had detected the most distant emission lines to date.

So Coe decided to be a one-person PR agent and spread the word himself. Throughout the rest of the conference he approached other cosmologists, holding out his phone to show them Larson's graph (while Larson, if she happened to be nearby, cringed because it didn't meet her personal standards). *Look,* he told them. *The most distant emission lines anyone has ever seen in the spectrum of a galaxy.*

This time Coe got the response he wanted, and he got it again and again: *Wow.*

The record didn't hold for long. No surprise; Coe had

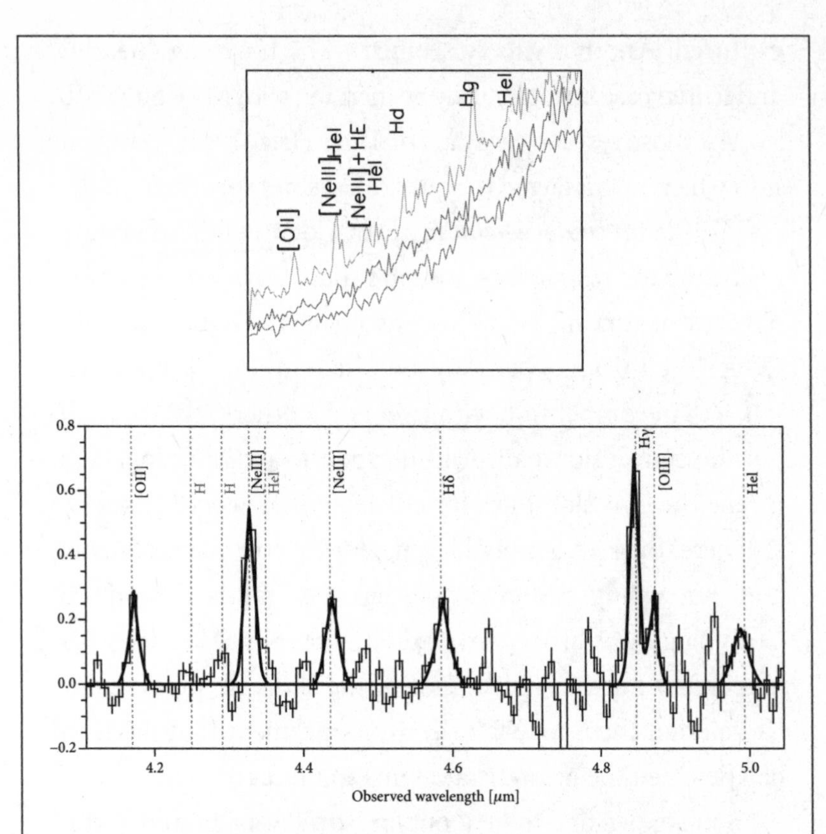

Before and after: [Top] Rebecca Larson's detections of elements in a galaxy approximately 460 million years after the Big Bang. The spectrum includes star light from the distant galaxy as well as zodiacal dust from our solar system. [Bottom] The final version as it appeared in the published paper led by Dan Coe's graduate student Tiger Hsiao and postdoc Abdurro'uf, after the team had subtracted the star light and zodiacal dust so that the elemental signatures would be easier to analyze. In both versions, the labels identify the elements and compounds corresponding to various peaks in the electromagnetic spectrum: OII and OIII are ionized oxygen; NeIII is ionized neon; He is helium; H is hydrogen; and so on.

expected the record to be broken soon. What he hadn't expected was what the emission lines in galaxy GNz-11 contained: nitrogen.

When Coe heard that news, he thought of the physicist Isidor Isaac Rabi's quip upon learning of the 1936 discovery of the muon, a particle possessing the same properties as an electron but a mass 207 times greater: "Who ordered that?"

Hydrogen, hydrogen, everywhere.

Specifically, single protons—the minimum requirement in order for a quantum entity to qualify as an element. Electrons existed then, too, in the immediate post–Big Bang period, and they would have bonded with the protons if photons hadn't been scattering them away. Instead, the hydrogen atoms remained ionized—meaning they had an electric charge owing to an imbalance in the number of positive protons (in the case of hydrogen, one) and negative electrons (none). By the time the universe was about 379,000 years old, however, the expansion of space had sapped the photons of enough energy that they were no longer any match for the electrons. At that moment single electrons began attaching themselves to single protons, neutralizing the previous imbalance, while the photons ran free across space and time—and continue to run free, carrying with them the imprint of that moment: the cosmic microwave background that two Bell Labs astronomers had detected in the 1960s.

Then the universe went what cosmologists call "dark." (Actually, it was flooding with photons, but they had nothing to illuminate.) And it stayed dark until, perhaps 100 million or 200 million years after the Big Bang, the neutral hydrogen began to coalesce into the first stars and galaxies, a process that in turn reionized the hydrogen. Hence the official title for Webb's fourth and final goal: "The End of the Dark Ages: First Light and Reionization."

When Webb launched on December 25, 2021, the universe was divided in five: Earth; solar system; the stars and exoplanets in our galaxy; the hundreds of billions of galaxies beyond our own; and, as Alan Dressler said in 1995, our origins—the reconfiguration of the quantum components of hydrogen atoms that triggered the first supernovae, which in turn initiated the creation cycle of heavier and heavier elements. Webb, in short, would be testing our understanding of the universe at the most fundamental levels.

Certainly many of Webb's observations in the early universe turned out to be the kind that cosmologists had hoped to find: a collection of primitive galaxies, for instance, that computer simulations showed would eventually evolve into a structure strikingly similar to the modern-day Coma Cluster, which contains at least a thousand galaxies—precisely the kind of result that can further science's understanding of galactic evolution.

But the presence of nitrogen in a galaxy only 440 million years after the Big Bang—and therefore after only one or two or three hundred million years of supernova seeding—was not what anyone expected. More to the point, it was not what the current models of galaxy evolution could explain.

Then again, a *lot* of what Webb was discovering in the early universe fell under the heading of "Who ordered that?" Right from the start—among the first observations that Webb made in the summer of 2022—early-universe researchers were reporting galaxies too large, too bright, too young to accommodate the current modeling. They were reporting black holes too massive and supernovae too numerous. The question, then, wasn't unreasonable: Had Webb broken the universe?

Yet the vast majority of cosmologists looked around and asked, *What's so broken?*

The instrument was an advance on previous technology, and it was producing data that was of far higher quality than these scientists had ever encountered. And the data was internally consistent, again and again. What it was inconsistent with was the existing models for galactic evolution in the early universe. As Ori Fox had heard repeatedly from the two holdouts on his team regarding the detection of supernova dust ejecta: A higher quality of data requires a higher quality of models. And in fact theorists adjusted their models sufficiently to explain some of those anomalies.

That's not breakage. That's science.

Sometimes the seeming anomalies that Webb was detecting didn't withstand follow-up observations of the kind that the scientific method demands. For instance, several initial estimates of the ages of mature galaxies at extremely high redshifts (and therefore at improbably early stages of development) disappeared under further spectroscopic analysis.

Sometimes anomalies turned out to have alternative explanations. Superbright galaxies turned out to have not "too many" stars but a supermassive black hole at their centers.

Sometimes, as was the case with those supermassive black holes, anomalies turned out to match a preexisting theory. Years earlier, Priyamvada Natarajan, a Yale astrophysicist and a member of Harvard's Black Hole Initiative, had suggested a growth mechanism involving gas aggregations that not only would have led to the formation of massive black holes early in the universe but would explain their prevalence in galaxies today. Her prediction, however, would lack matching evidence until the arrival of a powerful enough telescope — which Webb turned out to be.*

And sometimes anomalies led to the creation of new theories — a category to which the detection of an overabundance of nitrogen in GN-z11 belonged. One possible explanation of the overabundance had been elemental outpourings from a single star some 50,000 to 100,000 times more massive than the Sun. But in 2024 astronomers found that a theoretical model supposing two "bursts" of rapid star formation approximately 100 million years apart would create the

* This match of prediction and evidence won Natarajan a spot on the *Time* 100 list of the most influential people of 2024.

conditions that astronomers were observing. That result, in turn, supported a theory that "bursty star formation" explains the abundance of bright galaxies at an early stage in the universe's development.

Even so, cosmologists did have to *consider* the possibility that the Standard Model of Cosmology was broken. Considering the possibility that something somewhere is wrong is, according to the scientific method, the responsible approach when predictions and observations part ways. In the case of the Standard Model of Cosmology, considering the possibility that something somewhere was wrong wasn't just responsible; it was unavoidable. The Standard Model itself, after all, had arisen out of a series of discoveries that defied the prevailing logic at the time. The model, in fact, was the product of a procession of instances when predictions and observations had parted ways.

In the 1970s astronomers had found evidence in spiral galaxy after spiral galaxy that the rotation rates of the disks were too fast, at least according to prevailing theory: The outer edges of the galaxies were spinning as fast as the inner regions, seemingly in defiance of Newton's laws; it was as if, in our solar system, far-off Neptune was whipping around the Sun at the same rate as close-in Mercury. But when theorists posited that the galaxies were embedded in a sphere of matter that was invisible in any regime of the electromagnetic spectrum, the motions suddenly made sense. Since then, as improvements to telescopes allowed cosmologists to examine the universe on the whole, they found that the predictions emerging from the theoretical presence of this mysterious "dark matter" matched the observations of cosmic structures on the largest scales.

In the 1990s two teams of observers plotted their own versions of a Hubble diagram, the *x/y* graph suggesting that galaxies are receding at a rate proportional to their distance—the farther, the faster. The members of the two collaborations had assumed that in a universe full of matter that is gravitationally attracting all other matter, the expansion of space would be slowing. But by how much? Just enough that the expansion would eventually come to an eternal standstill? Or so much that the expansion would eventually reverse itself in a kind of about-face big bang?

Those teams used Cepheid variables as standard candles, just as Hubble had, but they also extended the "distance ladder" across the universe by adding to the category of standard candles a particular class of supernovae. Since Hubble's original diagram had indicated a straight-line correlation of velocity and distance, the two teams in the 1990s assumed that at some point that line would have to deviate from its 45-degree beeline rigidity, bending downward to indicate that distant objects were brighter and therefore nearer than one might otherwise expect.

By the first week of 1998 both collaborations had found evidence that the line did indeed diverge from 45 degrees—but again, prediction and observation had parted ways.

Instead of curving down, the line was curving up, indicating that the supernovae were *dimmer* than the observers expected and therefore that the expansion wasn't decelerating but *accelerating.*

The conclusion was as counterintuitive as any in the history of science: Earth isn't flat; Earth is not the center of the universe. Yet the community accepted it with alacrity: It made the universe add up.

In the late 1970s and early 1980s, theorists found that, according to quantum mechanics, the universe likely underwent an "inflation" that started about 10^{-36} second after the Big Bang (that is, at the fraction of a second that begins with a decimal point and ends 35 zeros and a 1 later) and finished, give or take, 10^{-33} second after the Big Bang. In the interim the universe increased its size by a factor of 10^{26}. Inflation thereby would have "smoothed out" space so that the universe would look roughly the same in all directions, as it does for us, no matter where we are in it. In scientific terms, the universe should be flat. And a flat universe dictates that the ratio between its actual mass-energy density and the density necessary to keep it from collapsing should be 1.

Before 1998, observations had indicated that the composition of the universe was nowhere near this critical density. It was maybe a third of the way there. But if you take the dark matter we can't see, add the stuff we can see (the protons and neutrons that make the universe accessible throughout the electromagnetic spectrum), and then add this new component

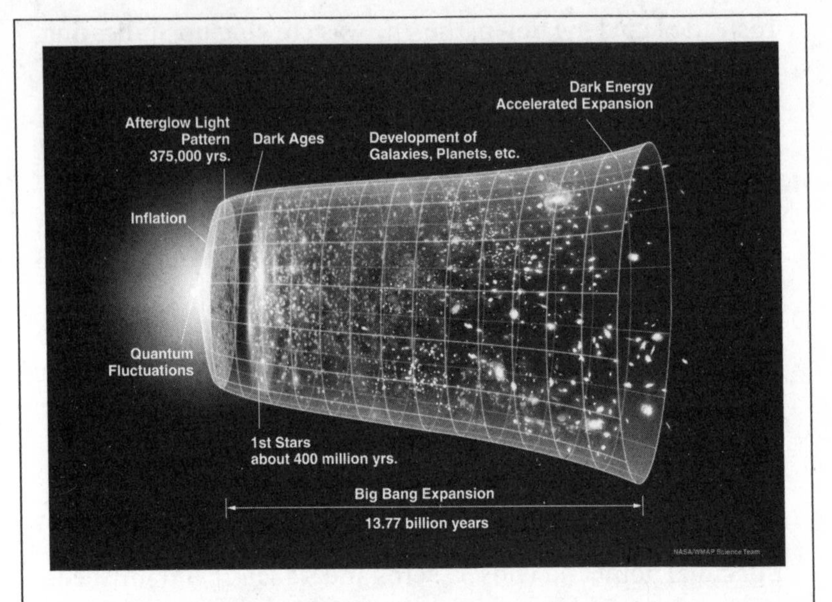

The Standard Model: Cosmologists conceive of the history of the universe not as an expansion *into* space but as the expansion *of* space, and they further conceive of the expansion of space as the thing we call *time*. According to the current interpretation of that history, the universe arose out of a quantum fluctuation. That fluctuation led to a state of "inflation" that, within the first fraction of a fraction of a second of the universe's existence, increased the size of the universe by a factor of 10^{26}. Some 379,000 years later the universe had cooled enough that matter and radiation decoupled from each other and went their separate ways, leaving an indelible imprint on the universe: the cosmic microwave background. Then the universe entered what scientists call the "dark ages." Webb's ultimate goal is to observe what came next: the first objects to possess the property that we identify with light—the stars and galaxies that glowed. Dark matter then exerted its gravitational influence, and dark energy exerted its "antigravitational" influence, on the growth of the universe at levels both cosmic (the overall structure) and intimate (you and me).

of "dark energy," the mass-energy density would exactly equal the density necessary to keep the universe from collapsing.

Einstein's lambda was back—and upon its return, the Standard Model of Cosmology was complete.

Yet it was also full of questions.

What is dark matter?

What is dark energy?

How much is dark energy speeding up the expansion of the universe? While cosmologists have agreed on the rate at which the expansion of the universe is accelerating *over time*, they disagree on the rate at which it's expanding *now*. Two methods of measurement have consistently produced two different answers outside each other's margin of error. Webb is probing the "crisis," as some cosmologists call it, by extending the distance ladder, but early results failed to resolve it.

And why does the universe add up, anyway? Change any of the values—the contributions of dark matter, dark energy, and regular matter to the overall mass-energy density of the universe—and the quantum properties that attended the Big Bang wouldn't have matured into a universe suitable for amoebas or galaxies, let alone life forms like ourselves. Depending on those values, the universe would have re-collapsed instantly or ballooned exponentially.

Yet those values, however improbably precise, are what kept cosmologists confident of the integrity of the Standard Model. The supporting evidence for the model was everywhere, literally.

One way to calculate the constitution of the universe is by studying the cosmic microwave background, the all-sky relic radiation dating to when the universe was only 379,000 years old. Take that picture and then compare it with simulations of literally millions of universes, each with its own amounts of regular matter, dark matter, and dark energy. Hypothetical universes with no regular matter or dark matter and 100 percent dark energy, or with 100 percent regular matter and no dark matter or dark energy, or with any combination in between, will all produce unique patterns.

The Wilkinson Microwave Anisotropy Probe, which launched in 2001 and delivered data on the cosmic microwave background from 2003 to 2012, provided one such census. Planck, an even more precise space observatory, began collecting its own cosmic microwave background data in 2009 and released its final results in 2018, corroborating the Wilkinson Probe's findings: The universe is 4.9 percent the stuff of us, 26.6 percent dark matter, and 68.5 percent dark energy.

When Webb cosmologists gathered at conferences, they expressed confidence in the Standard Model. The *models* within the model, however, might need adjusting.

Maybe a fundamental assumption was wrong—for instance, the relationship between the brightness of an early galaxy and its mass. When Edwin Hubble had created his diagram plotting nebula distances against velocities, he had erred in assuming that the luminosities of the island universes directly corresponded to their distances. In the end, though, the correspondence between luminosity and distance wound up being, in the aggregate, sufficient *enough*. Maybe the current assumption about the correlation between galactic mass and brightness in the early universe *wasn't* sufficient enough.

Then again, maybe initial conditions weren't quite what theorists had assumed. For instance, dust—that great variable in the interpretation of all observations galactic. What if supernovae eject dust in greater quantities than theorists had thought? Such a discrepancy could account for earlier formation of galactic structures.

Or, then again, so could a different distribution of dark matter in the early universe. Multiple Webb observations showed that once galactic evolution was well underway, in the first billion years after the Big Bang, it matched computer simulations incorporating the consensus values for dark matter. But during the first stages of galactic evolution, even a slight difference in the distribution of dark matter would have had an enormous gravitational effect: the collapse of gas and dust into stars, supernovae, galaxies, black holes.

Or maybe the influence of dark energy changes over time—a reasonable alternative, considering that cosmologists don't know what dark energy is, let alone how it works.

Or what if the existing particle census of the universe is off? Most scientists working on Webb were old enough to remember another imbalance between observation and theory: the "solar neutrino problem," a decades-long dispute about the abundance of a certain class of a fundamental particle, the neutrino, being emitted by the Sun. Theorists predicted one amount; neutrino detectors indicated another. Theorists suspected systematic errors in the observations. Astronomers questioned the completeness of the theory. Eventually theorists adjusted their own Standard Model, the one for particle physics, to allow neutrinos to have mass. A similar adjustment now in the Standard Model of Particle Physics — for instance, a new variety of neutrino in the early universe — might alter the distribution of mass and energy just enough to account for Webb's seemingly anomalous detections.

Or maybe the go-to answer for nearly a century applies here, too: We need a new physics. Only a decade after Einstein introduced his general theory of relativity, in 1915, quantum mechanics compromised its completeness. The universe of the very large — the one operating according to the rules of general relativity — proved to be mathematically incompatible with the universe of the very small — the one operating according to the rules of quantum mechanics.

But what if Webb did wind up breaking the Standard Model of Cosmology? Would that be so bad? What astronomer wouldn't want to be present at the creation of a new model? To be alive during Galileo's time, or William Herschel's, or Edwin Hubble's?

Even if Webb's scientists didn't get to break their own Standard Model, they couldn't believe their good fortune.

They got to be alive during the age of Webb.

EPILOGUE

In 1972, it was the telescope that would become Hubble, which eventually launched in 1990.

In 1982, it was the observatory that would become Chandra, which eventually launched in 1999.

In 1991, it was the telescope that would become Spitzer, which eventually launched in 2003.

In 2001, it was the telescope that would become Webb, which eventually launched in 2021.

In 2010, it was the telescope that would become Roman, which would eventually launch, maybe, in 2027.

And as of November 4, 2021, it was the Habitable Worlds Observatory, which was its name for now, anyway, and which wouldn't launch until, maybe, the early 2040s.

The setting was again the auditorium of the Space Telescope Science Institute. The occasion was the 2023 spring symposium, "Planetary Systems and the Origins of Life in the Era of JWST," and the speaker was the director of NASA's Astrophysics Division, Mark Clampin. He explained that he had chosen to open his talk with a slide showing the covers of the Astronomy and Astrophysics decadal surveys for a reason: He wanted to issue a reminder.

Yes, the science that Webb was performing was exemplary, especially so within the context of the current symposium—exoplanets. Clampin showed the Webb spectrum from WASP39-b, the first molecular and chemical profile of an exoplanet's atmosphere. It was one of those gasp-worthy graphs, the peaks and troughs from wavelengths 0.5 microns through 5.5 microns offering at-a-glance evidence for the presence of water, sodium, carbon monoxide, carbon dioxide, and sulfur dioxide.

Then he showed a Webb photograph of a protoplanetary system, complete with labels of its equally at-a-glance components: halo, outer ring, outer gap, intermediate belt, inner gap, and inner disk. He called out to a scientist in the audience who years earlier had created a model for what astronomers hoped Webb might identify in a protoplanetary disk: "I'm still astonished as to how close your model looks to what we have here"—a match between prediction and observation that was a credit to both the theorist and the instrument.

So, yes, they should all celebrate Webb, he said. But, Clampin added, they should also begin to turn their attention to the Habitable Worlds Observatory, even though its launch was still at least two decades away.

"It's a long haul," he said. "We're back to where we were with Webb, where we're really starting from scratch." But, he emphasized, they *were* starting—right now and right here.

This afternoon. This auditorium.

"It's you guys out here," he said, "the next generation of exoplanet scientists, who are going to make it happen."

As for the current generation of astronomers—the Webb generation that was now graying into its role as the previous generation—many of them convened four months later for the conference on the first year of Webb science. Among them was Mike Menzel. Sitting outside the cafeteria one lunchtime, in a little outdoor courtyard, he ran into Rogier Windhorst, a longtime colleague he hadn't seen in a few years. Windhorst was an early-universe specialist at Arizona State University and, like Heidi Hammel, one of Webb's six foundational interdisciplinary scientists. He and Menzel had been there from the start—or at least the post–2001 Decadal Survey start.

Windhorst hailed Menzel as the "Da Vinci of Webb"—the engineer who had hoarded the margins that had allowed for efficiencies that would keep the telescope operational for longer than anyone had previously had any right to hope. Longer even than the twenty years Webb officials were voicing publicly as a triumph. Twenty-three years, maybe. Maybe—*sotto voce*—longer.

Windhorst wondered if they would be around then. He and Menzel were both in their mid-sixties. Windhorst brightened: Maybe Menzel could be part of the Habitable Worlds Observatory team. Wouldn't he want to be part of the next generation, too?

Windhorst went inside then, leaving the question hanging in the autumnal air. The trees canopying the cafeteria terrace and the whole of the embankment leading down to Stony Run were starting to turn. Menzel finished his lunch. Then Windhorst reappeared.

Do you want to see something? he said to Menzel.

Of course.

Menzel accompanied Windhorst through the cafeteria, past the entrance to the auditorium, across the Institute's lobby, and into a conference room. Along the way they were joined by Marc Postman, the advisor who, in 2011, had counseled Dan Coe to proceed with caution when Coe said he'd discovered the most distant galaxy to date. On the conference table were waiting poster-sized printouts of Webb images. One of them was the Webb Deep Field, the image that Menzel's wife had said looked like the Hubble Deep Field. If her first look at the Webb Deep Field had been at this scale of magnification, though, she might not have made that mistake.

Then a couple of other Institute regulars—a postdoc, a publicist. Soon, under the low ceiling in a small room two floors down from where Riccardo Giacconi first asked Garth Illingworth to start thinking about a successor to Hubble *right now* and *right here,* one floor above the improbably still-humming Hubble mission control, one floor below Webb's Mission Operations Center, they were all bending over the Webb Deep Field image, pointing out the signature light arcs signifying that foreground galaxies were gravitationally lensing the infant galaxies in the universe.

Menzel then blasphemed.

The Habitable Worlds Observatory was all well and good. And sure, he could understand the scientific rationale for searching for exoplanets and their signs of life. He was a scientist, after all.

But—

And right then, right there, Mike Menzel voiced the kind of wild human yearning that Dan "Faster, Better, Cheaper" Goldin had tried to corral: Menzel said he wished that the Habitable Worlds Observatory would do something other than observe habitable worlds.

He said he wanted the next generation's space telescope to see even deeper into the origins of the universe.

He wanted more space.

He wanted more time.

He wanted more answers.

The Habitable Worlds Observatory would provide answers about exoplanets. If Menzel could have his way, though, it would also provide answers about the early universe. And if other astronomers could have their way over the next few years, it would probably provide answers to questions nobody had thought to ask. Almost assuredly it would provide those answers, if history was any guide. The specific answers, of course, had changed over the course of history—generation after generation, century after century. Once the answer had been more moons orbiting our sibling planets. Then the answer had been more sibling planets orbiting our star. Then the answer had been more stars in our galaxy. Then it was more galaxies, as far as we could see.

The question, however, never varied.

It was always the same.

It was *What's next?*

APPENDIX

The June 2023 issue of *Publications of the Astronomical Society of the Pacific* included a paper by hundreds of Webb veterans that represented their best attempt to preserve the collective several-decade effort of their own and their peers. It began, "We summarize the history, concept, scientific program, and technical performance of the James Webb Space Telescope." Following are the authors and their affiliations.

The James Webb Space Telescope Mission

Jonathan P. Gardner[1], John C. Mather[1], Randy Abbott[87,2], James S. Abell[1], Mark Abernathy[3], Faith E. Abney[3], John G. Abraham[1], Roberto Abraham[4,5], Yasin M. Abul-Huda[3], Scott Acton[2], Cynthia K. Adams[1], Evan Adams[3], David S. Adler[3], Maarten Adriaensen[6], Jonathan Albert Aguilar[3], Mansoor Ahmed[87,1], Nasif S. Ahmed[3], Tanjira Ahmed[1], Rüdeger Albat[6], Loïc Albert[7], Stacey Alberts[8], David Aldridge[9], Mary Marsha Allen[3], Shaune S. Allen[1], Martin Altenburg[10], Serhat Altunc[1], Jose Lorenzo Alvarez[11], Javier Álvarez-Márquez[12], Catarina Alves de

Oliveira[13], Leslie L. Ambrose[1], Satya M. Anandakrishnan[14], Gregory C. Andersen[1], Harry James Anderson[3], Jay Anderson[3], Kristen Anderson[14], Sara M. Anderson[3], Julio Aprea[6], Benita J. Archer[1], Jonathan W. Arenberg[14], Ioannis Argyriou[15], Santiago Arribas[12], Étienne Artigau[7], Amanda Rose Arvai[3], Paul Atcheson[87,2], Charles B. Atkinson[14], Jesse Averbukh[3], Cagatay Aymergen[1], John J. Bacinski[3], Wayne E. Baggett[3], Giorgio Bagnasco[11], Lynn L. Baker[1], Vicki Ann Balzano[3], Kimberly A. Banks[1], David A. Baran[1], Elizabeth A. Barker[3], Larry K. Barrett[1], Bruce O. Barringer[3], Allison Barto[2], William Bast[3], Pierre Baudoz[16], Stefi Baum[17], Thomas G. Beatty[18], Mathilde Beaulieu[19], Kathryn Bechtold[3], Tracy Beck[3], Megan M. Beddard[3], Charles Beichman[20], Larry Bellagama[14], Pierre Bely[87,3], Timothy W. Berger[14], Louis E. Bergeron[3], Antoine-Darveau Bernier[7], Maria D. Bertch[3], Charlotte Beskow[6], Laura E. Betz[1], Carl P. Biagetti[3], Stephan Birkmann[21], Kurt F. Bjorklund[14], James D. Blackwood[1], Ronald Paul Blazek[3], Stephen Blossfeld[14], Marcel Bluth[22], Anthony Boccaletti[16], Martin E. Boegner Jr[3], Ralph C. Bohlin[3], John Joseph Boia[3], Torsten Böker[21], N. Bonaventura[23], Nicholas A. Bond[1,24], Kari Ann Bosley[3], Rene A. Boucarut[1], Patrice Bouchet[25], Jeroen Bouwman[26], Gary Bower[3], Ariel S. Bowers[3], Charles W. Bowers[1], Leslye A. Boyce[1], Christine T. Boyer[3], Martha L. Boyer[3], Michael Boyer[3], Robert Boyer[3], Larry D. Bradley[3], Gregory R. Brady[3], Bernhard R. Brandl[27], Judith L. Brannen[1], David Breda[28], Harold G. Bremmer[87,1], David Brennan[3], Pamela A. Bresnahan[3], Stacey N. Bright[3], Brian J. Broiles[1], Asa Bromenschenkel[3], Brian H. Brooks[3], Keira J. Brooks[3], Bob Brown[87,2], Bruce Brown[14], Thomas M. Brown[3], Barry W. Bruce[87,1], Jonathan G. Bryson[1], Edwin D. Bujanda[14], Blake M. Bullock[14], A. J. Bunker[29], Rafael Bureo[11], Irving J. Burt[1], James Aaron Bush[3], Howard A. Bushouse[3], Marie C. Bussman[1], Olivier Cabaud[6], Steven Cale[1], Charles D. Calhoon[1], Humberto Calvani[3], Alicia M. Canipe[3], Francis M. Caputo[3], Mihai Cara[3], Larkin Carey[2], Michael Eli Case[3], Thaddeus Cesari[1], Lee D. Cetorelli[87,1], Don R. Chance[3], Lynn Chandler[1], Dave Chaney[2], George N. Chapman[3], S. Charlot[30], Pierre Chayer[3], Jeffrey I. Cheezum[14], Bin Chen[3], Christine H. Chen[3], Brian Cherinka[3], Sarah C. Chichester[3], Zachary S. Chilton[3], Dharini Chittiraibalan[3], Mark Clampin[31], Charles R. Clark[1], Kerry W. Clark[3], Stephanie M. Clark[1], Edward E. Claybrooks[1], Keith A. Cleveland[1], Andrew L. Cohen[14], Lester M. Cohen[32], Knicole D. Colón[1], Benee L. Coleman[3], Luis Colina[12], Brian J. Comber[1], Thomas M.

Comeau[3], Thomas Comer[3], Alain Conde Reis[6], Dennis C. Connolly[1], Kyle E. Conroy[3], Adam R. Contos[2,33], James Contreras[2], Neil J. Cook[7], James L. Cooper[1], Rachel Aviva Cooper[3], Michael F. Correia[1], Matteo Correnti[3], Christophe Cossou[34], Brian F. Costanza[14], Alain Coulais[35], Colin R. Cox[3], Ray T. Coyle[14], Misty M. Cracraft[3], Keith A. Crew[3], Gary J. Curtis[3], Bianca Cusveller[11], Cleyciane Da Costa Maciel[36], Christopher T. Dailey[1], Frédéric Daugeron[6], Greg S. Davidson[14], James E. Davies[3], Katherine Anne Davis[3], Michael S. Davis[1], Ratna Day[1], Daniel de Chambure[6,36], Pauline de Jong[36,11], Guido De Marchi[11], Bruce H. Dean[1], John E. Decker[87,1], Amy S. Delisa[1], Lawrence C. Dell[1], Gail Dellagatta[87,1], Franciszka Dembinska[6], Sandor Demosthenes[2], Nadezhda M. Dencheva[3], Philippe Deneu[37], William W. DePriest[3], Jeremy Deschenes[3], Nathalie Dethienne[37], Örs Hunor Detre[26], Rosa Izela Diaz[3], Daniel Dicken[38], Audrey S. DiFelice[3], Matthew Dillman[3], Maureen O. Disharoon[1], William V. Dixon[3], Jesse B. Doggett[3], Keisha L. Dominguez[1], Thomas S. Donaldson[3], Cristina M. Doria-Warner[1], Tony Dos Santos[36], Heather Doty[2], Robert E. Douglas, Jr[3], René Doyon[7], Alan Dressler[39], Jennifer Driggers[1], Phillip A. Driggers[1], Jamie L. Dunn[1], Kimberly C. DuPrie[3], Jean Dupuis[40], John Durning[87,1], Sanghamitra B. Dutta[31], Nicholas M. Earl[3], Paul Eccleston[41], Pascal Ecobichon[37], Eiichi Egami[8], Ralf Ehrenwinkler[10], Jonathan D. Eisenhamer[3], Michael Eisenhower[32], Daniel J. Eisenstein[32], Zaky El Hamel[11], Michelle L. Elie[3], James Elliott[3], Kyle Wesley Elliott[3], Michael Engesser[3], Néstor Espinoza[3], Odessa Etienne[3], Mireya Etxaluze[41], Leah Evans[3], Luce Fabreguettes[6], Massimo Falcolini[11], Patrick R. Falini[3], Curtis Fatig[87,1], Matthew Feeney[3], Lee D. Feinberg[1], Raymond Fels[11], Nazma Ferdous[3], Henry C. Ferguson[3], Laura Ferrarese[42], Marie-Héléne Ferreira[36], Pierre Ferruit[11,13], Malcolm Ferry[43], Joseph Charles Filippazzo[3], Daniel Firre[44], Mees Fix[3], Nicolas Flagey[3], Kathryn A. Flanagan[3], Scott W. Fleming[3], Michael Florian[8], James R. Flynn[14], Luca Foiadelli[44], Mark R. Fontaine[87,1], Erin Marie Fontanella[3], Peter Randolph Forshay[3], Elizabeth A. Fortner[87,1], Ori D. Fox[3], Alexandro P. Framarini[3], John I. Francisco[14], Randy Franck[2], Marijn Franx[27], David E. Franz[1], Scott D. Friedman[3], Katheryn E. Friend[14], James R. Frost[1], Henry Fu[14], Alexander W. Fullerton[3], Lionel Gaillard[11], Sergey Galkin[3], Ben Gallagher[2,45], Anthony D. Galyer[1], Macarena García Marín[21], Lisa E. Gardner[3], Dennis Garland[3], Bruce Albert Garrett[3], Danny Gasman[15], András Gáspár[8], René Gastaud[25], Daniel Gaudreau[40],

Peter Timothy Gauthier[3], Vincent Geers[38], Paul H. Geithner[1], Mario Gennaro[3], John Gerber[87,2], John C. Gereau[14], Robert Giampaoli[14], Giovanna Giardino[21], Paul C. Gibbons[1], Karoline Gilbert[3], Larry Gilman[14], Julien H. Girard[3], Mark E. Giuliano[3], Konstantinos Gkountis[6], Alistair Glasse[38], Kirk Zachary Glassmire[3], Adrian Michael Glauser[46], Stuart D. Glazer[1], Joshua Goldberg[3], David A. Golimowski[3], Shireen P. Gonzaga[3], Karl D. Gordon[3], Shawn J. Gordon[14], Paul Goudfrooij[3], Michael J. Gough[3], Adrian J. Graham[11], Christopher M. Grau[1], Joel David Green[3], Gretchen R. Greene[3], Thomas P. Greene[47], Perry E. Greenfield[3], Matthew A. Greenhouse[1], Thomas R. Greve[48], Edgar M. Greville[1], Stefano Grimaldi[2], Frank E. Groe[14], Andrew Groebner[3], David M. Grumm[3], Timothy Grundy[41], Manuel Güdel[49], Pierre Guillard[30], John Guldalian[14], Christopher A. Gunn[1], Anthony Gurule[2], Irvin Meyer Gutman[3], Paul D. Guy[88,1], Benjamin Guyot[6], Warren J. Hack[3], Peter Haderlein[28], James B. Hagan[3], Andria Hagedorn[14], Kevin Hainline[8], Craig Haley[9], Maryam Hami[3], Forrest Clifford Hamilton[3], Jeffrey Hammann[14], Heidi B. Hammel[50], Christopher J. Hanley[3], Carl August Hansen[3], Bruce Hardy[87,2], Bernd Harnisch[87,11], Michael Hunter Harr[3], Pamela Harris[1], Jessica Ann Hart[3], George F. Hartig[3], Hashima Hasan[31], Kathleen Marie Hashim[3], Ryan Hashimoto[14], Sujee J. Haskins[1], Robert Edward Hawkins[88,3], Brian Hayden[3], William L. Hayden[87,1], Mike Healy[11], Karen Hecht[3], Vince J. Heeg[14], Reem Hejal[14], Kristopher A. Helm[14], Nicholas J. Hengemihle[1], Thomas Henning[26], Alaina Henry[3], Ronald L. Henry[3], Katherine Henshaw[3], Scarlin Hernandez[3], Donald C. Herrington[3], Astrid Heske[11], Brigette Emily Hesman[3], David L. Hickey[3], Bryan N. Hilbert[3], Dean C. Hines[3], Michael R. Hinz[14], Michael Hirsch[14], Robert S. Hitcho[3], Klaus Hodapp[51], Philip E. Hodge[3], Melissa Hoffman[3], Sherie T. Holfeltz[3], Bryan Jason Holler[3], Jennifer Rose Hoppa[3], Scott Horner[47], Joseph M. Howard[1], Richard J. Howard[87,31], Jean M. Huber[1], Joseph S. Hunkeler[3], Alexander Hunter[3], David Gavin Hunter[3], Spencer W. Hurd[1], Brendan J. Hurst[3], John B. Hutchings[42], Jason E. Hylan[1], Luminita Ilinca Ignat[40], Garth Illingworth[52], Sandra M. Irish[1], John C. Isaacs III[3], Wallace C. Jackson Jr[14], Daniel T. Jaffe[53], Jasmin Jahic[14], Amir Jahromi[1], Peter Jakobsen[23], Bryan James[1], John C. James[1], LeAndrea Rae James[3], William Brian Jamieson[3], Raymond D. Jandra[14], Ray Jayawardhana[54], Robert Jedrzejewski[3], Basil S. Jeffers[1], Peter Jensen[11], Egges Joanne[87,2], Alan T. Johns[1], Carl A. Johnson[3], Eric L. Johnson[1],

Patricia Johnson[87,1], Phillip Stephen Johnson[3], Thomas K. Johnson[1], Timothy W. Johnson[3], Doug Johnstone[42,55], Delphine Jollet[11], Danny P. Jones[3], Gregory S. Jones[14], Olivia C. Jones[38], Ronald A. Jones[1], Vicki Jones[3], Ian J. Jordan[3], Margaret E. Jordan[3], Reginald Jue[14], Mark H. Jurkowski[1], Grant Justis[3], Kay Justtanont[56], Catherine C. Kaleida[3], Jason S. Kalirai[57], Phillip Cabrales Kalmanson[3], Lisa Kaltenegger[54], Jens Kammerer[3], Samuel K. Kan[14], Graham Childs Kanarek[3], Shaw-Hong Kao[3], Diane M. Karakla[3], Hermann Karl[10], Susan A. Kassin[3,58], David D. Kauffman[3], Patrick Kavanagh[59], Leigh L. Kelley[1], Douglas M. Kelly[8], Sarah Kendrew[21], Herbert V. Kennedy[3], Deborah A. Kenny[3], Ritva A. Keski-Kuha[1], Charles D. Keyes[3], Ali Khan[11], Richard C. Kidwell[3], Randy A. Kimble[1], James S. King[87,1], Richard C. King[1], Wayne M. Kinzel[3], Jeffrey R. Kirk[1], Marc E. Kirkpatrick[14], Pamela Klaassen[38], Lana Klingemann[2], Paul U. Klintworth[14], Bryan Adam Knapp[3], Scott Knight[2], Perry J. Knollenberg[14], Daniel Mark Knutsen[3], Robert Koehler[3], Anton M. Koekemoer[3], Earl T. Kofler[14], Vicki L. Kontson[1], Aiden Rose Kovacs[3], Vera Kozhurina-Platais[3], Oliver Krause[26], Gerard A. Kriss[3], John Krist[28], Monica R. Kristoffersen[14], Claudia Krogel[1], Anthony P. Krueger[3], Bernard A. Kulp[3], Nimisha Kumari[21], Sandy W. Kwan[28], Mark Kyprianou[3], Aurora Gadiano Labador[3], Álvaro Labiano[60], David Lafrenière[7], Pierre-Olivier Lagage[34], Victoria G. Laidler[3], Benoit Laine[11], Simon Laird[11], Charles-Philippe Lajoie[3], Matthew D. Lallo[3], May Yen Lam[3], Stephanie Marie LaMassa[3], Scott D. Lambros[1], Richard Joseph Lampenfield[3], Matthew Ed Lander[1], James Hutton Langston[3], Kirsten Larson[21], Melora Larson[28], Robert Joseph LaVerghetta[3], David R. Law[3], Jon F. Lawrence[1], David W. Lee[14], Janice Lee[3,8,61], Yat-Ning Paul Lee[3], Jarron Leisenring[8], Michael Dunlap Leveille[3], Nancy A. Levenson[3], Joshua S. Levi[14], Marie B. Levine[28], Dan Lewis[43], Jake Lewis[2,62], Nikole Lewis[54], Mattia Libralato[21], Norbert Lidon[37], Paula Louisa Liebrecht[3], Paul Lightsey[87,2], Simon Lilly[46], Frederick C. Lim[1], Pey Lian Lim[3], Sai-Kwong Ling[14], Lisa J. Link[1], Miranda Nicole Link[3], Jamie L. Lipinski[3], XiaoLi Liu[3], Amy S. Lo[14], Lynette Lobmeyer[2], Ryan M. Logue[3], Chris A. Long[3], Douglas R. Long[3], Ilana D. Long[3], Knox S. Long[3], Marcos López-Caniego[63], Jennifer M. Lotz[3], Jennifer M. Love-Pruitt[14], Michael Lubskiy[3], Edward B. Luers[87,1], Robert A. Luetgens[14], Annetta J. Luevano[14], Sarah Marie G. Flores Lui[3], James M. Lund III[14], Ray A. Lundquist[31], Jonathan Lunine[54], Nora Lützgendorf[21], Richard J. Lynch[1,64],

Alex J. MacDonald[3], Kenneth MacDonald[3], Matthew J. Macias[14], Keith I. Macklis[14], Peiman Maghami[1], Rishabh Y. Maharaja[1], Roberto Maiolino[65,66], Konstantinos G. Makrygiannis[14], Sunita Giri Malla[3], Eliot M. Malumuth[1], Elena Manjavacas[21], Andrea Marini[11], Amanda Marrione[3], Anthony Marston[13], André R Martel[3], Didier Martin[11], Peter G. Martin[67], Kristin L. Martinez[2], Marc Maschmann[10], Gregory L. Masci[3], Margaret E. Masetti[1,24], Michael Maszkiewicz[40], Gary Matthews[1], Jacob E. Matuskey[3], Glen A. McBrayer[14], Donald W. McCarthy[8], Mark J. McCaughrean[11], Leslie A. McClare[1], Michael D. McClare[1], John C. McCloskey[1], Taylore D. McClurg[14], Martin McCoy[1], Michael W. McElwain[1], Roy D. McGregor[14], Douglas B. McGuffey[1], Andrew G. McKay[14], William K. McKenzie[1], Brian McLean[3], Matthew McMaster[3], Warren McNeil[87,1], Wim De Meester[15], Kimberly L. Mehalick[1], Margaret Meixner[3], Marcio Meléndez[3], Michael P. Menzel[1], Michael T. Menzel[1], Matthew Merz[3], David D. Mesterharm[1], Michael R. Meyer[68], Michele L. Meyett[3], Luis E. Meza[14], Calvin Midwinter[9], Stefanie N. Milam[1], Jay Todd Miller[3], William C. Miller[1], Cherie L. Miskey[1], Karl Misselt[8], Eileen P. Mitchell[1], Martin Mohan[14], Emily E. Montoya[1], Michael J. Moran[14], Takahiro Morishita[3], Amaya Moro-Martín[3], Debra L. Morrison[3], Jane Morrison[8], Ernie C. Morse[3], Michael Moschos[14], S. H. Moseley[1,69], Gary E. Mosier[1], Peter Mosner[10], Matt Mountain[50], Jason S. Muckenthaler[14], Donald G. Mueller[3], Migo Mueller[70], Daniella Muhiem[88,1], Prisca Mühlmann[11], Susan Elizabeth Mullally[3], Stephanie M. Mullen[1], Alan J Munger[14], Jess Murphy[2], Katherine T. Murray[3], James C. Muzerolle[3], Matthew Mycroft[28], Andrew Myers[3], Carey R. Myers[3], Fred Richard R. Myers[14], Richard Myers[14], Kaila Myrick[3], Adrian F. Nagle, IV[2], Omnarayani Nayak[3], Bret Naylor[28], Susan G. Neff[1], Edmund P. Nelan[3], John Nella[14], Duy Tuong Nguyen[3], Michael N. Nguyen[1], Bryony Nickson[3], John Joseph Nidhiry[3], Malcolm B. Niedner[87,1], Maria Nieto-Santisteban[3], Nikolay K. Nikolov[3], Mary Ann Nishisaka[14], Alberto Noriega-Crespo[3], Antonella Nota[87,21], Robyn C. O'Mara[1], Michael Oboryshko[3], Marcus B. O'Brien[14], William R. Ochs[87,1], Joel D. Offenberg[71,72], Patrick Michael Ogle[3], Raymond G. Ohl[1], Joseph Hamden Olmsted[3], Shannon Barbara Osborne[3], Brian Patrick O'Shaughnessy[3], Göran Östlin[73], Brian O'Sullivan[21], O. Justin Otor[3], Richard Ottens[1], Nathalie N.-Q. Ouellette[7], Daria J. Outlaw[1], Beverly A. Owens[3], Camilla Pacifici[3], James Christophe Page[3], James G. Paranilam[3], Sang Park[32], Keith A. Parrish[1], Laura

Paschal[1], Polychronis Patapis[46], Jignasha Patel[1], Keith Patrick[14], Robert A. Pattishall Jr[14], Douglas William Paul[3], Shirley J. Paul[1], Tyler Andrew Pauly[3], Cheryl M. Pavlovsky[3], Maria Peña-Guerrero[3], Andrew H. Pedder[3], Matthew Weldon Peek[3], Patricia A. Pelham[3], Konstantin Penanen[28], Beth A. Perriello[3], Marshall D. Perrin[3], Richard F. Perrine[3], Chuck Perrygo[87,1], Muriel Peslier[36], Michael Petach[14], Karla A. Peterson[3], Tom Pfarr[87,1], James M. Pierson[1], Martin Pietraszkiewicz[14], Guy Pilchen[6], Judy L. Pipher[74], Norbert Pirzkal[21], Joseph T. Pitman[1], Danielle M. Player[3], Rachel Plesha[3], Anja Plitzke[11], John A. Pohner[14], Karyn Konstantin Poletis[3], Joseph A. Pollizzi[3], Ethan Polster[3], James T. Pontius[1], Klaus Pontoppidan[3], Susana C. Porges[14], Gregg D. Potter[14], Stephen Prescott[3], Charles R. Proffitt[3], Laurent Pueyo[3], Irma Aracely Quispe Neira[3], Armando Radich[87,1], Reiko T. Rager[3], Julien Rameau[7,75], Deborah D. Ramey[88,1], Rafael Ramos Alarcon[3], Riccardo Rampini[11], Robert Rapp[1], Robert A. Rashford[1], Bernard J. Rauscher[1], Swara Ravindranath[3], Timothy Rawle[21], Tynika N. Rawlings[1], Tom Ray[59], Michael W. Regan[3], Brian Rehm[87,1], Kenneth D. Rehm[76], Neill Reid[3], Carl A. Reis[1], Florian Renk[44], Tom B. Reoch[14], Michael Ressler[28], Armin W. Rest[3], Paul J. Reynolds[14], Joel G. Richon[3], Karen V. Richon[1], Michael Ridgaway[3], Adric Richard Riedel[3], George H. Rieke[8], Marcia J. Rieke[8], Richard E. Rifelli[14], Jane R. Rigby[1], Catherine S. Riggs[3], Nancy J. Ringel[1], Christine E. Ritchie[3], Hans-Walter Rix[26], Massimo Robberto[3,58], Gregory L. Robinson[87,31], Michael S. Robinson[3], Orion Robinson[3], Frank W. Rock[3], David R. Rodriguez[3], Bruno Rodríguez del Pino[12], Thomas Roellig[47], Scott O. Rohrbach[1], Anthony J. Roman[3], Frederick J. Romelfanger[3], Felipe P. Romo Jr[1], Jose J. Rosales[1], Perry Rose[3], Anthony F. Roteliuk[14], Marc N. Roth[14], Braden Quinn Rothwell[3], Sylvain Rouzaud[37], Jason Rowe[77], Neil Rowlands[9], Arpita Roy[3], Pierre Royer[15], Chunlei Rui[14], Peter Rumler[87,11], William Rumpl[3], Melissa L. Russ[3], Michael B. Ryan[14], Richard M. Ryan[31], Karl Saad[40], Modhumita Sabata[3], Rick Sabatino[1], Elena Sabbi[3], Phillip A. Sabelhaus[88,1], Stephen Sabia[1], Kailash C. Sahu[3], Babak N. Saif[1,3], Jean-Christophe Salvignol[11], Piyal Samara-Ratna[78], Bridget S. Samuelson[14], Felicia A. Sanders[28], Bradley Sappington[3], B. A. Sargent[3,58], Arne Sauer[10], Bruce J. Savadkin[87,1], Marcin Sawicki[79], Tina M. Schappell[1], Caroline Scheffer[11], Silvia Scheithauer[26], Ron Scherer[14], Conrad Schiff[1], Everett Schlawin[8], Olivier Schmeitzky[11], Tyler S. Schmitz[3], Donald J. Schmude[14], Analyn Schneider[28], Jürgen

Schreiber[26], Hilde Schroeven-Deceuninck[11], John J. Schultz[3], Ryan Schwab[3], Curtis H. Schwartz[1], Dario Scoccimarro[6], John F. Scott[3], Michelle B. Scott[1], Bonita L. Seaton[1], Bruce S. Seely[3], Bernard Seery[80], Mark Seidleck[87,1], Kenneth Sembach[3], Clare Elizabeth Shanahan[3], Bryan Shaughnessy[41], Richard A. Shaw[3], Christopher Michael Shay[3], Even Sheehan[1], Kartik Sheth[31], Hsin-Yi Shih[3], Irene Shivaei[8], Noah Siegel[2], Matthew G. Sienkiewicz[3], Debra D. Simmons[14], Bernard P. Simon[3], Marco Sirianni[21], Anand Sivaramakrishnan[3,58,81], Jeffrey E. Slade[1], G. C. Sloan[3], Christine E. Slocum[3], Steven E. Slowinski[3], Corbett T. Smith[1], Eric P. Smith[31], Erin C. Smith[1], Koby Smith[2], Robert Smith[82], Stephanie J. Smith[3], John L. Smolik[14], David R. Soderblom[3], Sangmo Tony Sohn[3], Jeff Sokol[2], George Sonneborn[87,1], Christopher D. Sontag[3], Peter R. Sooy[1], Remi Soummer[3], Dana M. Southwood[14], Kay Spain[3], Joseph Sparmo[1], David T. Speer[1], Richard Spencer[3], Joseph D. Sprofera[14], Scott S. Stallcup[3], Marcia K. Stanley[1], John A. Stansberry[3], Christopher C. Stark[1], Carl W. Starr[1], Diane Y. Stassi[1], Jane A. Steck[1], Christine D. Steeley[1], Matthew A. Stephens[1], Ralph J. Stephenson[14], Alphonso C. Stewart[1], Massimo Stiavelli[3], Hervey Stockman Jr[87,3], Paolo Strada[11], Amber N. Straughn[1], Scott Streetman[2], David Kendal Strickland[3], Jingping F. Strobele[14], Martin Stuhlinger[13], Jeffrey Edward Stys[3], Miguel Such[11], Kalyani Sukhatme[28], Joseph F. Sullivan[87,2], Pamela C. Sullivan[1], Sandra M. Sumner[1], Fengwu Sun[8], Benjamin Dale Sunnquist[3], Daryl Allen Swade[3], Michael S. Swam[3], Diane F. Swenton[1], Robby A. Swoish[14], Oi In Tam Litten[3], Laszlo Tamas[38], Andrew Tao[14], David K. Taylor[3], Joanna M. Taylor[3], Maurice te Plate[21], Mason Van Tea[3], Kelly K. Teague[3], Randal C. Telfer[3], Tea Temim[83], Scott C. Texter[14], Deepashri G. Thatte[3], Christopher Lee Thompson[3], Linda M. Thompson[3], Shaun R. Thomson[1], Harley Thronson[87,1], C. M. Tierney[14], Tuomo Tikkanen[78], Lee Tinnin[8], William Thomas Tippet[3], Connor William Todd[3], Hien D. Tran[3], John Trauger[28], Edwin Gregorio Trejo[3], Justin Hoang Vinh Truong[3], Christine L. Tsukamoto[14], Yasir Tufail[3], Jason Tumlinson[3], Samuel Tustain[41], Harrison Tyra[3], Leonardo Ubeda[3], Kelli Underwood[3], Michael A. Uzzo[3], Steven Vaclavik[3], Frida Valenduc[36], Jeff A. Valenti[3], Julie Van Campen[1], Inge van de Wetering[11], Roeland P. Van Der Marel[3], Remy van Haarlem[11], Bart Vandenbussche[15], Ewine F. van Dishoeck[27], Dona D. Vanterpool[1], Michael R. Vernoy[14], Maria Begoña Vila Costas[1,22], Kevin Volk[3], Piet Voorzaat[11], Mark F. Voyton[1], Ekaterina Vydra[3], Darryl J.

Waddy[1], Christoffel Waelkens[15], Glenn Michael Wahlgren[3], Frederick E. Walker Jr[14], Michel Wander[40], Christine K. Warfield[3], Gerald Warner[9], Francis C. Wasiak[1], Matthew F. Wasiak[1], James Wehner[14], Kevin R. Weiler[14], Mark Weilert[28], Stanley B. Weiss[14], Martyn Wells[38], Alan D. Welty[3], Lauren Wheate[1], Thomas P. Wheeler[3], Christy L. White[14], Paul Whitehouse[1], Jennifer Margaret Whiteleather[3], William Russell Whitman[3], Christina C. Williams[84], Christopher N. A. Willmer[8], Chris J. Willott[42], Scott P. Willoughby[14], Andrew Wilson[9], Debra Wilson[8], Donna V. Wilson[1], Rogier Windhorst[85], Emily Christine Wislowski[3], David J. Wolfe[3], Michael A. Wolfe[3], Schuyler Wolff[8], Amancio Wondel[36], Cindy Woo[14], Robert T. Woods[14], Elaine Worden[87,2], William Workman[3], Gillian S. Wright[38], Carl Wu[1], Chi-Rai Wu[3], Dakin D. Wun[14], Kristen B. Wymer[3], Thomas Yadetie[3], Isabelle C. Yan[1], Keith C. Yang[14], Kayla L. Yates[3], Christopher R. Yeager[3], Ethan John Yerger[3], Erick T. Young[80], Gary Young[14], Gene Yu[14], Susan Yu[3], Dean S. Zak[3], Peter Zeidler[86], Robert Zepp[3], Julia Zhou[9], Christian A. Zincke[1], Stephanie Zonak[3], and Elisabeth Zondag[11]

[1] NASA Goddard Space Flight Center, 8800 Greenbelt Road, Greenbelt, MD 20771, USA; jonathan.p.gardner@nasa.gov

[2] Ball Aerospace & Technologies Corp., 1600 Commerce Street, Boulder, CO 80301, USA

[3] Space Telescope Science Institute, 3700 San Martin Drive, Baltimore, MD, 21218, USA

[4] Department of Astronomy & Astrophysics, University of Toronto, 50 Saint George Street, Toronto, ON M5S 3H4, Canada

[5] Dunlap Institute for Astronomy and Astrophysics, University of Toronto, 50 Saint George Street, Toronto, ON M5S 3H4, Canada

[6] European Space Agency, HQ Daumesnil, 52 rue Jacques Hillairet, F-75012 Paris, France

[7] Institut de Recherche sur les Exoplanètes (iREx), Université de Montréal, Département de Physique, C.P. 6128 Succursale Centre-ville, Montréal, QC H3C 3J7, Canada

[8] Steward Observatory, University of Arizona, 933 North Cherry Avenue, Tucson, AZ 85721, USA

[9] Honeywell Aerospace #100, 303 Terry Fox Drive, Ottawa, ON K2K 3J1, Canada

[10] Airbus Defence and Space GmbH, Ottobrunn, Germany

[11] European Space Agency, European Research & Technology Centre, Keplerlaan 1, Postbus 299, 2200 AG Noordwijk, The Netherlands
[12] Centro de Astrobiología (CAB, CSIC-INTA), Carretera de Ajalvir, E-28850 Torrejón de Ardoz, Madrid, Spain
[13] European Space Agency, European Space Astronomy Centre, Camino Bajo, del Castillo, s/n, Urbanización Villafranca del Castillo, E-28692 Villanueva de la Cañada, Madrid, Spain
[14] Northrop Grumman, One Space Park, Redondo Beach, CA 90278, USA
[15] Instituut voor Sterrenkunde, KU Leuven, Celestijnenlaan 200D, Bus-2410, B-3000 Leuven, Belgium
[16] LESIA, Observatoire de Paris, Université PSL, CNRS, Sorbonne Université, Université de Paris, 5 place Jules Janssen, F-92195 Meudon, France
[17] Faculty of Science, 230 Machray Hall, 186 Dysart Road, University of Manitoba, Winnipeg, MB R3T 2N2, Canada
[18] Department of Astronomy, University of Wisconsin-Madison, Madison, WI 53706, USA
[19] Université Côte d'Azur, Observatoire de la Côte d'Azur, CNRS, Laboratoire Lagrange, F-06108 Nice, France
[20] NASA Exoplanet Science Institute/IPAC, Jet Propulsion Laboratory, California Institute of Technology, 1200 East California Boulevard, Pasadena, CA 91125, USA
[21] European Space Agency, Space Telescope Science Institute, 3700 San Martin Drive, Baltimore, MD 21218, USA
[22] KBR, 7701 Greenbelt Road, Greenbelt, MD 20770, USA
[23] Cosmic Dawn Center (DAWN), Niels Bohr Institute, University of Copenhagen, Jagtvej 128, DK-2200, Denmark
[24] Adnet Systems, Inc., 6720B Rockledge Drive, Suite # 504, Bethesda, MD 20817, USA
[25] Laboratoire AIM Paris-Saclay, CEA-IRFU/SAp, CNRS, Université Paris Diderot, F-91191 Gif-sur-Yvette, France
[26] Max Planck Institute for Astronomy, Königstuhl 17, D-69117 Heidelberg, Germany
[27] Leiden Observatory, Leiden University, PO Box 9513, 2300 RA Leiden, The Netherlands
[28] Jet Propulsion Laboratory, California Institute of Technology, 4800 Oak Grove Drive, Pasadena, CA 91109, USA
[29] Department of Physics, University of Oxford, Denys Wilkinson Building, Keble Road, Oxford, OX1 3RH, UK
[30] Sorbonne Université, UPMC-CNRS, UMR7095, Institut d'Astrophysique de Paris, F-75014 Paris, France
[31] NASA Headquarters, 300 E Street SW, Washington, DC 20546, USA

[32] The Center for Astrophysics, 60 Garden Street, Cambridge, MA 02138, USA
[33] Moog Space and Defense Group, 5025 North Robb Street, Suite 500, Arvada, CO 80033, USA
[34] Université Paris-Saclay, Université de Paris, CEA, CNRS, AIM, F-91191 Gif-sur-Yvette, France
[35] LERMA (CNRS) & Observatoire de Paris, Paris, France
[36] European Space Agency, Centre Spatial Guyanais, BP816 — Route Nationale 1, 97388 Kourou CEDEX, French Guiana, France
[37] Centre national d'études spatiales, Direction des Lanceurs, 52 rue Jacques Hillairet, F-75612 Paris CEDEX, France
[38] UK Astronomy Technology Centre, Royal Observatory Edinburgh, Blackford Hill, Edinburgh, EH9 3HJ, UK
[39] The Observatories, The Carnegie Institution for Science, 813 Santa Barbara Street, Pasadena, CA 91101, USA
[40] Canadian Space Agency, 6767 Route de l'Aéroport, Saint-Hubert, QC J3Y 8Y9, Canada
[41] RAL Space, STFC, Rutherford Appleton Laboratory, Harwell, Oxford, Didcot, OX11 0QX, UK
[42] NRC Herzberg, 5071 West Saanich Road, Victoria, BC V9E 2E7, Canada
[43] Lockheed Martin Advanced Technology Center, 3251 Hanover Street, Palo Alto, CA 94304, USA
[44] European Space Agency, European Space Operations Centre, Robert-Bosch-Strasse 5, D-64293 Darmstadt, Germany
[45] TMT International Observatory, 100 W. Walnut Street, Suite 300, Pasadena, CA, 91124, USA
[46] ETH Zurich, Wolfgang-Pauli-Str 27, CH-8093 Zurich, Switzerland
[47] NASA Ames Research Center, Space Science and Astrobiology Division, MS 245-6, Moffett Field, CA, 94035, USA
[48] DTU Space, Technical University of Denmark. Building 328, Elektrovej, DK-2800 Kgs. Lyngby, Denmark
[49] Dept. of Astrophysics, University of Vienna, Türkenschanzstr 17, A-1180 Vienna, Austria
[50] Associated Universities for Research in Astronomy, Inc., 1331 Pennsylvania Avenue Northwest, Suite 1475, Washington, DC 20004, USA
[51] Institute for Astronomy, 640 North Aohoku Place, Hilo, HI 96720, USA
[52] UCO/Lick Observatory, University of California, Santa Cruz, CA 95064, USA
[53] The University of Texas at Austin, Department of Astronomy RLM 16.342, Austin, TX 78712, USA
[54] Department of Astronomy, Cornell University, Ithaca, NY 14853, USA

[55] Department of Physics and Astronomy, University of Victoria, Victoria, BC, V8P 5C2, Canada
[56] Dept. of Space, Earth and Environment, Chalmers University of Technology, Onsala Space Observatory, SE-43992 Onsala, Sweden
[57] Johns Hopkins University Applied Physics Laboratory, 11100 Johns Hopkins Road, Laurel, MD 20723, USA
[58] Dept. of Physics & Astronomy, Johns Hopkins University, 3400 North Charles Street, Baltimore, MD, 21218, USA
[59] Dublin Institute for Advanced Studies, School of Cosmic Physics, 31 Fitzwilliam Place, Dublin 2, D02 XF86, Ireland
[60] Telespazio UK for the European Space Agency, ESAC, Camino Bajo del Castillo s/n, E-28692 Villanueva de la Cañada, Spain
[61] Gemini Observatory/NSF's NOIRLab, 950 North Cherry Avenue, Tucson, AZ, 85719, USA
[62] Blue Canyon Technologies, 5330 Airport Road, Boulder, CO, 80301, USA
[63] Aurora Technology for the European Space Agency, ESAC, Madrid, Spain
[64] HelioSpace Inc., 932 Parker Street, Suite 2, Berkeley, CA, 94710, USA
[65] Cavendish Laboratory, University of Cambridge, 19 J.J. Thomson Avenue, Cambridge, CB3 0HE, UK
[66] Kavli Institute for Cosmology, University of Cambridge, Madingley Road, Cambridge, CB3 0HA, UK
[67] Canadian Institute for Theoretical Astrophysics, University of Toronto, McLennan Physical Laboratories, 60 Saint George Street, Toronto, Ontario, M5S 3H8, Canada 4 Publications of the Astronomical Society of the Pacific, 135:068001 (24pp), 2023 June Gardner et al.
[68] Astronomy Department, University of Michigan, Ann Arbor, MI 48109, USA
[69] Quantum Circuits, Inc., New Haven, Connecticut, USA
[70] Kapteyn Astronomical Institute, University of Groningen, P.O. Box 800, 9700 AV Groningen, The Netherlands
[71] Vantage Systems Inc, Greenbelt, MD 20706, USA
[72] Howard Community College, Columbia, MD 21044, USA
[73] Department of Astronomy, Oskar Klein Centre, Stockholm University, SE-106 91 Stockholm, Sweden
[74] Department of Physics and Astronomy, University of Rochester, Rochester, NY 14627, USA
[75] Univ. Grenoble Alpes, CNRS, IPAG, F-38000 Grenoble, France
[76] Katherine Johnson IV&V Facility, Goddard Space Flight Center, Code 180, Greenbelt, MD 20771, USA
[77] Department of Physics & Astronomy, Bishop's University, Sherbrooke, QC J1M 1Z7, Canada

[78] School of Physics & Astronomy, Space Research Centre, University of Leicester, Space Park Leicester, 92 Corporation Road, Leicester, LE4 5SP, UK
[79] Institute for Computational Astrophysics and Department of Astronomy & Physics, Saint Mary's University, 923 Robie Street, Halifax, NS B3H 3C3, Canada
[80] Universities Space Research Association, 425 3rd Street Southwest, Suite 950, Washington DC 20024, USA
[81] Astrophysics Department, American Museum of Natural History, 79th Street at Central Park West, New York, NY 10024, USA
[82] Department of History and Classics, University of Alberta, Edmonton, Alberta, Canada
[83] Princeton University, 4 Ivy Lane, Princeton, NJ 08544, USA
[84] National Optical-Infrared Research Laboratory, 950 North Cherry Avenue, Tucson, AZ 85719, USA
[85] School of Earth and Space Exploration, Arizona State University, Tempe, AZ 85287-1404, USA
[86] AURA for the European Space Agency (ESA), ESA Office, Space Telescope Science Institute, 3700 San Martin Drive, Baltimore, MD 21218, USA
[87] Retired
[88] Deceased

ACKNOWLEDGMENTS

Deep thanks to Bruce Nichols, whose vision inspired this book. Equally deep thanks to editor extraordinaire Alexander Littlefield, not least for guiding this mission through its own "six months of terror" (you have to read the book to get the reference). Yet more deep thanks to agent David Granger, a new colleague in possession of ancient wisdom. Why isn't Katya Rice in the Copyediting Hall of Fame?—and no, *Because it doesn't exist* isn't a good enough reason. Special thanks to Lee Billings, Ron Cowen, Clara Moskowitz, Nicholas Suntzeff, and Christopher Wanjek for advice and insight. Thanks also to the team at Little, Brown: Linda Arends, Kay Banning, Erin Cain, Bryan Christian, Allan Fallow, Deborah Jacobs, Pat Jalbert-Levine, Gregg Kulick, Gabrielle Leporati, Laura Mamelok, and Morgan Wu. Appreciation to the Faculty Development Fund at Goddard College for research support. And of course great gratitude to the book's interviewees, whose names are in the following Notes section. Names not in the Notes include hundreds of Webb scientists and experts whose talks at conferences, publications in journals, and casual asides in corridors or cafeterias were invaluable.

NOTES

Twenty thousand is the consensus number for how many people worked on the James Webb Space Telescope from its inception in the 1980s through its scientific commissioning in July 2022. One hundred million is the consensus number for how many hours they worked. Fifty gigabytes—50 billion units of individual meaning, whether a word or a number—is the consensus quantity for how much data Webb shovels Earthward every day. The science that has begun to emerge in the short time that Webb has been operational overwhelms even the scientists. Ask them what they think are the most important results to date even within their own narrow areas of specialization and the standard response is *I wouldn't know where to begin.*

All of which is to say that this book is not comprehensive. It is instead, I hope, representative: of the human investment of intellectual and physical (and emotional) labor over the past four decades, and of the science that Webb has been producing at a nearly incomprehensible rate and in a nearly indigestible volume.

In an attempt to make that material comprehensible and digestible, I have thought of each chapter as having three components. A brief description of each of those components might be helpful as a guide to navigating the notes that follow.

One component of each chapter involves the personal experiences of a scientist or several scientists. I have named those persons in **boldface**. (The same is true of other **interviewees**.)

A second component is the science. As with the scientist or scientists who provide a narrative throughline for each chapter, the science in each chapter is a sampling of the breakthroughs Webb has achieved. The specific results from that sampling merit formal citations in the following notes. But again, those results are suggestive of a much more extensive collective effort.

The third component of the chapters is a historical context—a brief history of the kind of science that has come to define cosmology today and therefore has also helped define Webb's technological and intellectual ambitions. Those sources you will not find below. They are too numerous to credit or, perhaps, remember; this component of the book is an amalgamation of my research over the course of several books. Having once listened to a journalist friend deliver a diatribe about authors who cite themselves, I prefer not to indulge in that practice myself, if only to avoid providing fodder for ridicule from writers in bars. Still, readers who are curious about the history and philosophy of astronomy, astrophysics, and cosmology might want to consult one or more of my other books.

Finally, for anyone wanting to explore Webb on their own, two official websites are essential: https://webbtelescope.org/home and https://www.stsci.edu/jwst. And for more information about the Space Telescope Science Institute, see https://www.stsci.edu/.

PROLOGUE

Opening: **Dan Coe**; **Rebecca Larson**.

To watch the Webb unveiling at the White House, see: https://www.youtube.com/watch?v=ySaIPoHisRg.

CHAPTER ONE: VISION

Success has many fathers, but failure is an orphan: Several interviewees offered variations on this maxim regarding rival post-launch claims as to the "real" origin of the telescope that would become Webb: the initial 1980s discussions or the 1990s reboot. **Garth Illingworth** and **Alan Dressler** offered distillations of the two predominant versions of the Webb origin story; what appears here adheres to the publicly available facts while identifying those two participants' interpretations.

For the Giacconi paper, see: Riccardo Giacconi et al., "Evidence for X-Rays from Sources Outside the Solar System," *Physical Review Letters* 9 (December 1, 1962).

For the Space Studies Board's final report, see: National Research Council, *Space Science in the Twenty-First Century: Imperatives for the Decades 1995 to 2015* (Washington, DC: National Academy Press, 1988).

For Illingworth's Baltimore address, see: D. McNally, ed., *Highlights of Astronomy 8* (Dordrecht, Netherlands: Kluwer Academic Publishers, 1989).

For more on the 1989 workshop at the Space Telescope Science Institute, see: Pierre-Yves Bely, Christopher J. Burrows, and Garth D. Illingworth, eds., *The Next Generation Space Telescope: Proceedings of a Workshop Held at the Space Telescope Science Institute, Baltimore, Maryland, 13–15 September 1989* (Baltimore: Space Telescope Science Institute, 1989).

For context on George H. W. Bush's remarks on returning to the moon, see: National Research Council of the National Academies, *NASA's Strategic Direction and the Need for a National Consensus* (Washington, DC: National Academies Press, 2012).

For more on infrared astronomy within the context of Webb, see: https://webbtelescope.org/webb-science/the-observatory/infrared-astronomy.

For an overview of the original 1995 Hubble Deep Field, see: https://hubblesite.org/contents/articles/hubble-deep-fields.

CHAPTER TWO: MISSION

The story of Dan Goldin addressing Alan Dressler at the January 1996 American Astronomical Society meeting is widely available. See, for example: David S. Leckrone, *Life with Hubble: An Insider's View of the World's Most Famous Telescope* (Bristol, UK: IOP Publishing, 2020). (Also: **Alan Dressler**.)

Background on the mission from about 1996 to post-launch: **Torsten Böker**; **Dan Coe**; **Alan Dressler**; **Ori Fox**; **Garth Illingworth**; **Mike Menzel**; **Brian O'Sullivan**; **Marcia J. Rieke**; **Massimo Stiavelli**.

For the HST & Beyond Committee report, see: Alan Dressler, ed., *HST and Beyond: Exploration and the Search for Origins: A Vision for Ultraviolet-Optical-Infrared Space Astronomy* (Washington, DC: Association of Universities for Research in Astronomy, 1996).

For more on the Science Working Group's conclusions, see: Jonathan P. Gardner et al., "The James Webb Space Telescope," *Space Science Reviews* 123 (2006).

For more on the state of Webb scientific planning through 2008, see: Harley A. Thronson, Massimo Stiavelli, and Alexander Tielens, eds., *Astrophysics in the Next Decade: The James Webb Space Telescope and Concurrent Facilities* (Dordrecht, Netherlands: Springer, 2009).

For more on the history and construction of Webb, see: Chris Gunn and Christopher Wanjek, *Inside the Star Factory: The Creation of the James Webb Space Telescope, NASA's Largest and Most Powerful Space Observatory* (Cambridge, MA: MIT Press, 2023).

Letter from Barbara A. Mikulski to Charles Bolden, June 29, 2010: courtesy of Garth Illingworth.

Letter from Charles F. Bolden, Jr., to the Honorable Barbara A. Mikulski, July 21, 2010: courtesy of Garth Illingworth.

"Terms of Reference (TOR) for Independent Comprehensive Review Panel of the James Webb Space Telescope, July 15, 2010": courtesy of Garth Illingworth.

For more on Webb's mid-crisis status as of October 2010, see: Lee Billings, "Space Science: The Telescope That Ate Astronomy," *Nature,* October 27, 2010.

For the "James Webb Space Telescope (JWST) Independent Comprehensive Review Panel (ICRP) Final Report," see: www.nasa.gov/wp-content/uploads/2015/01/499224main_jwst-icrp_report-final.pdf?emrc=da04aa.

For the "Independent Comprehensive Review Panel" cover letter of November 5, 2010, from John R. Casani to Administrator Charles Bolden, see: www.nasa.gov/wp-content/uploads/2015/01/499276main_casani_letter.pdf.

For "NASA's Detailed Response to the James Webb Space Telescope Independent Comprehensive Review Panel Report," see: www.webb.nasa.gov/resources/JamesWebbSpaceTelescopeIndependentComprehensiveReviewPanelReport.pdf.

For the February 2018 United States Government Accountability Office report to Congressional Committees, see: *James Webb Space Telescope Integration and Test Challenges Have Delayed Launch and Threaten to Push Costs over Cap*: https://www.gao.gov/assets/700/690693.pdf.

For on-the-ground coverage of the days before launch, see: Marina Koren, "The Most Exciting Spot in the Cosmos Right Now Is French Guiana," *The Atlantic,* December 23, 2021: https://www.theatlantic.com/science/archive/2021/12/james-webb-space-telescope-launch-french-guiana/621109/.

For more on the post-launch mood among Webb scientists, see: Dennis Overbye and Joey Roulette, "A Giant Telescope Grows in Space," *New York Times,* January 8, 2022.

CHAPTER THREE: FIRST HORIZON

Opening and throughout: **Heidi Hammel**.

Background on Webb's solar-system science: **Imke de Pater**; **Stefanie Millam**; **John Stansberry**; **Cristina Thomas**.

For more on Hammel, see: Fred Bortz, *Beyond Jupiter: The Story of Planetary Astronomer Heidi Hammel* (Washington, DC: Joseph Henry Press, 2005).

To watch the 1994 Comet Shoemaker-Levy 9 press conference, see: https://ntrs.nasa.gov/citations/19990116991.

For more on spectroscopy within the context of Webb, see: https://webbtelescope.org/contents/articles/spectroscopy-101—invisible-spectroscopy.

For more on the DART collision, see: Tereza Pultarova, "James Webb Space Telescope Pushed Past Its Limits to Observe DART Asteroid Crash," Space.com, February 8, 2023: https://www.space.com/dart-impact-forces-webb-through-limit.

Enceladus: **Geronimo Villanueva**.

For more on Enceladus, see both: Ron Cowen, "Giant Plume Spotted Erupting from Moon of Saturn Might Contain Ingredients for Life," *Science,* May 30, 2023: https://www.science.org/content/article/giant-plume-spotted-erupting-moon-saturn-might-contain-ingredients-life; and Alexandra Witze, "JWST Spots Biggest Water Plume Yet Spewing from a Moon of Saturn," *Nature,* May 18, 2023: https://www.nature.com/articles/d41586-023-01666-x.

CHAPTER FOUR: SECOND HORIZON

Opening and throughout: **Nikku Madhusudhan**.

Background on Webb's exoplanet science: **Néstor Espinoza**; **Nikku Madhusudhan**.

For more on the "First Year of JWST Science" conference, see: https://www.stsci.edu/contents/events/stsci/2023/september/the-first-year-of-jwst-science-conference.

For the Madhusudhan paper, see: Nikku Madhusudhan et al., "Carbon-Bearing Molecules in a Possible Hycean Atmosphere" (2023): https://arxiv.org/abs/2309.05566.

For more on the art and science of Webb imagery, see, well: "The Art and Science of Webb Imagery": https://www.youtube.com/watch?v=dJX0RAyuqos.

For more on L1527, see: Sarah Kuta, "James Webb Captures a Protostar in a Fiery Hourglass," smithsonianmag.com, November 17, 2022: https://www.smithsonianmag.com/smart-news/james-webbs-captures-a-protostar-in-a-fiery-hourglass-180981149/.

For more on the observation of water in a protoplanetary disk, see: Andrea Banzatti et al., "JWST Reveals Excess Cool Water Near the Snow Line in Compact Disks, Consistent with Pebble Drift," *Astrophysical Journal Letters* 957 (2023).

For more on the Chamaeleon I observations, see: M. K. McClure et al., "An Ice Age JWST Inventory of Dense Molecular Cloud Ices," *Nature Astronomy* 7 (2023).

For more on HIP 65426 b, see: Aarynn L. Carter et al., "The JWST Early Release Science Program for Direct Observations of Exoplanetary Systems I: High Contrast Imaging of the Exoplanet HIP 65426 b from 2–16 μm" (2023): https://arxiv.org/abs/2208.14990.

CHAPTER FIVE: THIRD HORIZON

Opening and throughout: **Ori Fox.**

Background on Webb's galactic science: **Richard Ellis; Ori Fox; Svea Hernandez; Nora Lützgendorf; Adam Riess.**

For more on the supernovae in NGC 6946, see: Melissa Shahbandeh et al., "JWST Observations of Dust Reservoirs in Type IIP Supernovae 2004et and 2017eaw," *Monthly Notices of the Royal Astronomical Society* 523 (2023).

For more on the PHANGS collaboration's early results, see: "PHANGS-JWST First Results," *Astrophysical Journal Letters* 944 (2023): https://iopscience.iop.org/collections/2041-8205_PHANGS-JWST-First-Results.

For more on observations of the neutron star at the center of 1987A, see both: Daniel Clery, "Stellar Remains of Famed 1987 Supernova Found at Last," *Science,* February 22, 2024; and C. Fransson et al., "Emission Lines Due to Ionizing Radiation from a Compact Object in the Remnant of Supernova 1987A," *Science* 383 (2024).

CHAPTER SIX: FINAL HORIZON

Opening and throughout: **Dan Coe; Rebecca Larson.**

Background on Webb's early-universe science: **Dan Coe; Andrey Kravtsov; Rebecca Larson; Mike Menzel; Rogier Windhorst.**

For more on the January 2023 American Astronomical Society meeting, see: https://aas.org/meetings/aas241.

For more on the Hubble Space Telescope observation of MACS0647-JD, see: Dan Coe et al., "CLASH: Three Strongly Lensed Images of a Candidate z ~ 11 Galaxy," *Astrophysical Journal* 762 (2012).

For more on the Webb observation of MACS0647-JD, see both: Tiger Yu-Yang Hsiao et al., "JWST Reveals a Possible z ~ 11 Galaxy Merger in Triply Lensed MACS0647–JD," *Astrophysical Journal Letters* 949 (2023); and Tiger Yu-Yang Hsiao et al., "JWST NIRSpec Spectroscopy of the Triply-lensed z = 10.17 galaxy MACS0647–JD," *Astrophysical Journal* accepted (2024).

For more on whether Webb broke cosmology, see: Rebecca Boyle, "No, the James Webb Space Telescope Hasn't Broken Cosmology," Wired.com, September 26, 2023: https://www

.wired.com/story/no-the-james-webb-space-telescope-hasnt-broken-cosmology/.

For more on the nitrogen in galaxy GN-z11, see: Chiaki Kobayashi and Andrea Ferrara, "Rapid Chemical Enrichment by Intermittent Star Formation in GN-z11," *Astrophysical Journal Letters* 962 (2024).

EPILOGUE

For more on the "Planetary Systems and the Origins of Life in the Era of JWST" symposium, see: https://www.stsci.edu/contents/events/stsci/2023/may/planetary-systems-and-the-origins-of-life-in-the-era-of-jwst?timeframe=past&timeframe=past&timeframe=past&timeframe=past&page=4&filterUUID=24ba27ed-9d32-4e90-aad8-3ee1a1784ae8.

ILLUSTRATION CREDITS

Color images

1 (Webb's First Deep Field): NASA, ESA, CSA, and STScI

2–3 (Jupiter): NASA, ESA, Jupiter ERS Team; image processing by Ricardo Hueso (UPV/EHU) and Judy Schmidt

4–5 (Star birth): ESA/Webb, NASA, CSA, T. Ray (Dublin Institute for Advanced Studies)

6–7 (Cosmic Cliffs): NASA, ESA, CSA, STScI

8, 9 (Pillars of Creation, Parts I and II): NASA, ESA, CSA, STScI; Joseph DePasquale (STScI), Anton M. Koekemoer (STScI), Alyssa Pagan (STScI)

10–11 (Dust): ESA/Webb, NASA & CSA, M. Meixner

12–13 (Spiral galaxies): NASA, ESA, CSA, STScI, Janice Lee (STScI), Thomas Williams (Oxford), PHANGS Team, Elizabeth Wheatley (STScI)

14 (Stephan's Quintet): NASA, ESA, CSA, and STScI

15 (Galactic merger): ESA/Webb, NASA & CSA, L. Armus, A. Evans; the Hubble Heritage Team (STScI/AURA)-ESA/Hubble Collaboration

16 (Gravitational lensing): ESA/Webb, NASA & CSA, J. Rigby and the JWST TEMPLATES team

Design of the color insert: Richard Panek

Black-and-white images

76: NASA/ESA/CSA

78, 79: NASA/STScI

112: NASA, ESA, CSA, STScI, and G. Villanueva (NASA's Goddard Space Flight Center). Image processing: A. Pagan (STScI).

116: NASA, ESA, CSA, STScI

135: NASA, ESA, CSA, STScI

160: NASA, ESA, CSA, Ori D. Fox (STScI), Michael Dulude (STScI)

178: Rebecca Larson (RIT / UT Austin); Tiger Yu-Yang Hsiao et al., *Astrophysical Journal*

186: NASA/WMAP Science Team

INDEX

Note: *Italic* page numbers refer to illustrations.

Index

Index

Index

Index

ABOUT THE AUTHOR

Richard Panek is the author of numerous books, including *The 4% Universe: Dark Matter, Dark Energy, and the Race to Discover the Rest of Reality,* for which he received the Science Communication Award from the American Institute of Physics. He has also been the recipient of a Guggenheim fellowship in science writing, among other honors. Panek and Temple Grandin collaborated on *The Autistic Brain,* a *New York Times* bestseller. His own books have been translated into sixteen languages, and he has contributed to publications including the *New York Times,* the *Washington Post, Scientific American, Smithsonian, Natural History,* and *Esquire.* He lives in New York City.